Ecoagriculture:

Initiatives in Eastern and Southern Africa

Ecoagriculture:

Initiatives in Eastern and Southern Africa

J.F. Devlin and T. Zettel, editors

UNIVERSITY of GUELPH

WEAVER PRESS

Published by:
Weaver Press, P O Box A1922, Avondale, Harare, Zimbabwe

Cover design: Danes Design, Harare

Layout: Fontline Electronic Publishing, Harare

Printed in Zimbabwe by Mazongororo Paper Converters.

ISBN 0 7974 2015 0

Environmental Capacity Enhancement Project

Environmental Roundtable Series

John Devlin, series editor

Table of Contents

Series Preface

The Environmental Capacity Enhancement Project (ECEP) has sought over the past four years to contribute to improved regional environmental management in Eastern and Southern Africa. The project has involved participants from 13 countries (Angola, Botswana, Kenya, Lesotho, Malawi, Mozambique, Namibia, South Africa, Swaziland, Tanzania, Uganda, Zambia, Zimbabwe) and has concentrated upon six environmental areas: eco-tourism, ecological agriculture, land use planning, small scale mining, waste management and watershed management. The project has not sought to develop a narrow set of regional environmental experts. Rather it has sought to bring together development practitioners, activists, analysts, researchers and policy-makers so that all could become familiar with the policy options and technology choices available to respond to the complex range of environmental problems shared across the region.

The regional training courses, the collaborative research projects, and the professional exchanges supported by ECEP have created multistakeholder venues where representatives from governmental and non-governmental institutions, the private sector and academic institutions could build linkages for mutual support and on-going information exchange. During its four years of operation ECEP has graduated 180 from its training courses and funded 53 collaborative research projects involving 125 principal researchers. It has also supported 35 professional enhancement attachments of which 27 have involved exchanges between different countries. Having brought environmental practitioners together ECEP then sought to establish the conditions for continued networking among them so that the regional and multistakeholder perspective would be maintained over the long term. Six networks of researchers and practitioners have been created. And all have had an opportunity to meet during the networking conference held in Harare from May 13-15, 1998 to exchange papers, discuss their future organizational efforts and evaluate what had been achieved to date.

The final outputs of the project are the five volumes of collected papers comprising the Environmental Roundtable Series. These five volumes represent the best of documents and reports that have been produced by ECEP participants. They will serve not only as valuable collections for environmental policy makers, environmental educators and environmental activists but also as catalysts for future initiatives dedicated to building an effective environmental management culture in the region. The preparation of this series has been made possible by two important groups. First, the Canadian International Development Agency, which has generously funded ECEP over the past four years. Second, the many ECEP participants who have generously donated their time and energy to making this project a success. Their contributions are acknowledged with thanks.

Dr. R.J. McLaughlin

Dean, Ontario Agricultural College, University of Guelph, Canada

March 1999

Preface

When we first spoke about the possibility of coordinating a training course on ecological agriculture to be held in Siavonga, Zambia, John's concern was to find someone who could speak to a variety of agricultural practitioners about the wide range of issues involved in ecological farming. Ted had gained a reputation as a reliable champion of ecological agriculture based on his own efforts to move his 100 ha. farm toward sustainability and on his involvement with grassroots training, organizing and lobbying to help the fledgling ecological farming movement in North America. But Ted was skeptical. It seemed he was being asked to speak to an agriculture that was distantly remote from his organic dairy farm in the snow belt of Southern Ontario. To Ted's surprise and delight he quickly realized that he had much in common with African farmers. The crises in African farming, the history of ill-fated attempts to ease hunger and expand opportunities, usually with inappropriate technologies and inadequate attention to ecological and cultural effects, were problems with which he was familiar. When a canvass for potential resource persons working in Eastern and Southern Africa resulted in an impressive list of practitioners already engaged in research and promotion of ecological agriculture in the region, what had originally been conceived as a training course evolved into a roundtable. The task became one of giving an opportunity for African agricultural practitioners to speak to each other. We were both very happy with this result.

The Ecological Agriculture Roundtable brought together a diverse mixture of some sixty individuals. The thirty presenters included Deans of Agricultural Schools, subsistence farmers, government agents, extension workers, leaders of farmers groups, community workers, and agricultural researchers. The thirty-odd "trainees" included an equally broad range of people active in agricultural production, training, research and extension. Their active participation broadened and deepened the discussions.

At the roundtable the problem for ecological agriculture was defined as two fold: first to produce enough food and second to devise an agriculture which does not threaten the long term health of the resource base. The leading phrase was "To

Feed the People Forever".

The sense of solidarity at the roundtable was palpable. We all recognized that farming is a task that has been dreadfully undervalued. But more importantly we recognized that farming must again become a vocation shared by farm families across the globe. All who struggle with the earth to wrestle from her our daily bread are allies. Different techniques are used according to time and place, but the partnership with the earth that consumes our energies and energizes our efforts is timeless and universal. It binds us together. We do not cultivate money, exploiting the land until we're rich enough to abandon it. We are part of the earth, cooperating with her for the purpose of succession. We do what we do in order that we will be able to keep on doing it, forever!

John F. Devlin
Ted Zettel

Contributors

A.D.J. Ambali is a professor in the Biology Department, Chancellor College, University of Malawi, P.O. Box 280, Zomba, Malawi. Tel:+265-522-222. Fax:+265-522-046. E-mail: aambali@unima.wn.apc.org

Margriet Berkhout is a visiting researcher at Veld Products Research, P.O. Box 2020, Gaborone, Botswana. Tel and Fax: +267-347047. E-mail: veldprod@info.bw

Rachel Bezner Kerr was a M.Sc. Student in the Department of Land Resource Science, University of Guelph. She graduated in 1998. Tel:+519-763-9789. E-mail: rachel@first-light.net

Randall F. Brummett was an associate of the SADC/ICLARM Aquaculture Network P.O. Box 229, Zomba, Malawi. He now is assigned to Cairo. Tel: +265 531 274, Fax:+265 522 733, E-mail: rbrummett@cgnet.com.

Fredson J.K Chikafumbwa was an associate of the SADC/ICLARM Aquaculture Network P.O. Box 229, Zomba, Malawi until his untimely death in 1998.

Sean Davison is a member of faculty at the Department of Microbiology, University of the Western Cape, P/Bag X17, Bellville, South Africa. Tel:+27-21-959-2216. Fax:+27-21-959-2266. E-mail:sean@mbiol.uwc.ac.za

Paul Desmarais, S.J. is an instructor at the Kasisi Agricultural Training Centre, Box 30652, Lusaka, Zambia. Tel: +260 01 233101, Fax: +260 01 293763, E-mail: katc@zamserv.zamtel.zm.

John F. Devlin is International Projects Coordinator for the School of Rural Planning and Development, University of Guelph, Guelph, Ontario, Canada, N1G 2W1. Tel:+1-519-824-4120 Ext.3773. Fax:+1-519-821-9904. E-mail:jdevlin@uoguelph.ca

Archileo N. Kaaya is a member of the faculty of the Department of Food Science and Technology, Makerere University, P.O. Box 7062, Kampala, Uganda. Tel:+256-41-533676. Fax:+256-41-531641. E-mail: fsptjku@arcc.or.ke

J. J. Kanjanga is Farm Director of the Lipangwe Organic Manure Demonstration Farm, P.O. Box 26, Matale, Ntcheu, Malawi. Fax: +265 671 427.

Peter Kevan is a member of the faculty of the Department of Environmental

Biology, University of Guelph, Guelph, Ontario, Canada, N1G 2W1. Tel: +519 824-4120 Ext.2479. Fax:+519 837-0442. E-mail: pkevan@uoguelph.ca

Sehaam Khan is a researcher at the at the Agricultural Research Council: Vegetable and Ornamental Plant Institute (Western Cape), Private Bag X1, Elsenburg 7607, South Africa. Tel:+27-21-8085416. Fax: +27-21-8085440.

Dumisani Kutywayo, Coffee Research Station, P.O. Box 61, Chipinge, Zimbabwe. Tel:+263-27- 2476. Fax:+263-27-2542. E-mail:rsscoffee@mango.zw

Clive Lightfoot is an associate of the International Support Group and Agropolis International, Avenue Agropolis, 34394 Montpellier CX5, France. Tel: +33 4 67047527. Fax: +33 4 67047526. E-mail: clive.lightfoot@agropolis.fr

Obed I.M. Lungu is a member of the Department of Soil Science, University of Zambia, P.O. Box 32379 Lusaka, E-mail: olungu@agric.unza.zm.

Zacharia J.U. Malley is a research associate of Marti Uyole, P.O.Box 400, Mbeya, Tanzania.

Alfred 0. Maluwa is a research associate of the National Aquaculture Centre, P.O. Box 44, Domasi, Malawi. Tel:+265-531-215. Fax:+265-522-397.

Livai C. Matarirano is Director of the Fambidzanai Permaculture Centre, Box CY 301, Causeway, Harare, Zimbabwe. Tel: +263 111 604434. Fax: +263 4 726911. E-mail: fambidzanai@mango.zw.

Stanley M. Mateke is Senior Horticultural Research Officer, Veld Products Research, P.O. Box 2020, Gaborone, Botswana. Tel and Fax: +267-347047. E-mail: veldprod@info.bw

Henry F. Mbeza is a Senior Lecturer in the Department of Processing and Storage Engineering, Bunda College of Agriculture, Box 219, Lilongwe, Malawi. Tel:+265-277-222. Fax:+265-277-364. E-mail:hfmbeza@unima.wn.apc.org

Abel M. Mbwile is an Agricultural Extension Officer with the Agricultural Development Trust Fund, P.O. Box 204, Mbozi, Tanzania. Fax:+255-65-4246.

Terence P. McGonigle, is a member of the Faculty of Agriculture, Gifu University, 1-1 Yanagido, Gifu-shi Gifu 501-11, Japan. E-mail: tmcgonig@cc.gifu-u.ac.jp

I. J. Minde is a Post-Doctoral Research Fellow in the Agriculture Policy Unit, Bunda College of Agriculture, Box 219, Lilongwe, Malawi. Tel:+265-277-222. Fax:+265-277-364.

Letlamoreng Mosenene was Director of the Machobane Agricultural Development Foundation. She is currently Natural Resources Management Programme Coordinator for Helvetas Lesotho. P.O. Box 708, Maseru 100, Lesotho. Tel/Fax:

+266 317875. E-mail: helvetas@lesoff.co.za.

Herbert K. Murwira is a researcher at the Tropical Soil Biology and Fertility Programme, c/o ACFD, Box A469, Avondale, Harare, Zimbabwe. Tel:+263-4-860421. Fax:+ 263-4-860423. E-mail: hmurwira@zambezi.net.

Sunday Mutabazi is an associate of ASTEX, African Science and Technology Exchange, Kampala, Uganda. Tel:+256-41-235193. Fax: +256-41-345597. E-mail: astex@imul.com.

Dezi S. Ngambeki is an agricultural consultant based in Kampala. AIMS Agribusiness Enterprises, P.O. Box 11321, Kampala, Uganda. E-mail: astex@imul.com

Reg P. Noble is a farming systems consultant now living in Canada. Tel: +1 416 544-0735. Fax: +1 416 440-2809. E-mail: regnoble@web.net.

Jean K. Nzuma is a Research Officer of the Chemistry and Soil Research Institute, Box CY550 Causeway, Harare, Zimbabwe. Tel:+263-4-704541 Ext.2068. Fax:+263-4-728317.

Yucabeth M. Ogega is an officer of the Ministry of Lands and Settlement, Nyamira District, Kenya.

John W. Potter is a nematologist with Agriculture & AgriFood Canada, Southern Crop Protection & Food Research Centre, 4902 Victoria Ave. N. Box 6000, Vineland Station, Ontario, Canada.. Tel: +905 562-4113 Ext.232, Fax: +905 562-4335, E-mail: potterj@em.agr.ca

Sarah Rosloski is an M.Sc. student in the Department of Environmental Biology, University of Guelph, Guelph, Ontario, Canada. E-mail:sroslosk@uoguelph.ca

Ramesh C. Saxena is an associate of the International Centre of Insect Physiology and Ecology (ICIPE), P. 0. Box 30772, Nairobi, Kenya. Fax:+254 385 22190. E-mail: rcsaxena@thorntree.org

Tim E. Simalenga is a member of the Faculty of Agriculture, University of Fort Hare, P/Bag X1314, Alice 5700, South Africa and the Chair of the Animal Traction Network for East and Southern Africa. (ATNESA). Tel: +27 404 22085. Fax: +27 404 31730. E-mail:tim.s@ufhcc.ufh.ac.za

Robert C. Sinclair is a faculty member of the Department of Botany, University of Pretoria, pretoria 0002, South Africa. Tel/Fax: +27 12 342-2783. E-mail: sinclair@scientia.up.ac.za

Peter van Straaten, is an Adjunct Professor in the Department of Land Resource Science, University of Guelph, Ontario, Canada, N1G 2W1. Tel: +519 824-

4120 ext. 2454. Fax:+ 519 824-5730. E-mail: pvanstra@lrs.uoguelph.ca.

Alex R. Tindimubona is Chairman of ASTEX, African Science and Technology Exchange, Kampala, Uganda. Tel:+256-41-235193. Fax: +256-41-345597. E-mail: astex@imul.com.

Mark G. Wright is a Research Entomologist at the Agricultural Research Council: Vegetable and Ornamental Plant Institute (Western Cape), Private Bag X1, Elsenburg 7607, South Africa. Tel:+27-21-8085416. Fax: +27-21-8085440. E-mail: mark@igs5.agric.za

Ted Zettel is a dairy farmer. RR#1, Chepstow, Ontario, Canada, N0G 1K0. Tel:+1-519-366-9982. Fax:+1-519-366-9917. E-mail:jervic@wcl.on.ca

INTRODUCTION

by John F. Devlin and Ted Zettel

The papers that are collected together here are products of two distinct programmes of the Environmental Capacity Enhancement Project. The majority of papers were presented by their authors at the Roundtable on Ecological Agriculture which was held in Siavonga, Zambia from the 28 May to the 4 June 1997. The sections of the book follow the agenda of the Ecological Agriculture Roundtable. However, not all of the thirty presentations made at the roundtable could be included. Some of the presentation were made from notes that were not written up into paper form and thus had to be left aside. Thus the collection serves as a partial proceedings for that event. The remaining papers are reports on research projects funded by ECEP. Some of these research papers were presented to the Ecological Agriculture Network meeting held in Harare in May of 1998.

All the papers share a concern to promote an ecological approach to farming. Ecological agriculture begins and ends with an appreciation for the connectedness of all living things, from the smallest micro-organism in the soil to the "crown of creation", humankind. Such an appreciation calls for an understanding of the many side effects of any intervention. It demands that we maintain a vision of the whole farm even while tackling the smallest detail. This whole farm includes not only livestock, crops, soil, water and air, but all the natural flora and fauna in and around the farm. In this complex and often delicate balance of interconnected systems the farmer must carefully place each step.

Successful farming begins with the living soil. The first and largest section of the book directs attention to a variety of techniques and approaches to soil management. Obed Lungu describes some traditional cropping systems in Zambia and how they interact to threaten or protect soil vitality. Zacharia Malley introduces a unique and effective system of pit cultivation used on the steep slopes in southwestern Tanzania. Ngambeki, Tindimubona and Mutabazi describe the current state of mountain farming in Central Kigezi, Uganda, and the decline of the terracing system that had maintained productivity there for many years. Letlamoreng Mosenene sets out the basic principles of the Machobane farming system developed in Lesotho. Timothy Simalenga then reports on his work in Lesotho studying the labour input of the Machobane system and some labour saving tools that were introduced to reduce labour input. The importance of manure to soil fertility is the focus of the next three papers. Murwira and Nzuma report on several studies on manure management undertaken in Zimbabwe. They compare above ground with below ground manure storage. Then Peter van Straaten reports on some preliminary work on the use of mineral additives to supplement manure quality. This is followed by Rachel Bezner Kerr's study on the use of manure extract on garden vegetables in Malawi. Soil management is important not only for crops but also for trees. The last two papers in this section look at issues relating to agroforestry. Margriet Berkhout describes her research on planting techniques for maximizing water capture in an agroforestry system in Botswana. Mateke, McGonigle and Sinclair then end the section with a description of three research initiatives studying the relationship between mycorrhizal fungi in the soil and the growth success of indigenous fruit trees in South Africa, Botswana and Zimbabwe.

Pest management is the area of agricultural management where the ecological farmer parts company most dramatically with the conventional farmer. Ecological farming seeks to drastically reduce or eliminate the use of synthetic pesticides. In the next section on integrated pest management Mark Wright offers an overview of IPM in South Africa. Ramesh Saxena describes some of the results of his work in Kenya with neem as a botanical pesticide. Rosloski, Kevan and Potter then describe some experiments in Namibia on the use of mulches to control root knot nematode. They also describe an associated training course conducted at the research site. Dumisani Kutywayo surveys the pests and some control techniques used on coffee in Zimbabwe. Abel Mbwile then surveys some common botanical controls he has found in Tanzania. Finally Wright, Khan and Davison describe their research on fungus and viral control agents for vegetable pests in South Africa.

The book then turns to the problem of conducting effective agricultural research for the small farmer. Lightfoot and Noble outline the evolution of the farming system approach to agricultural research and describe their approach with its emphasis on the impact of ecology on the study of farming systems. An important innovation in the ecological approach to farming systems has been the recognition of the critical role that fish ponds can play in integrating the use of farm resources. Crop and livestock waste can feed fish and fish waste can fertilize crops while the fish are harvested for human consumption and market. Chikafumbwa, Brummett and Noble set out the basics of this technology. Then Brummett and Chikafumbwa describe how research on fish pond development is being approached by ICLARM in Malawi. This section ends with a description of Maluwa and Ambali's work on the development of fish feedstock for fish ponds in Malawi.

These sections provide a variety of perspectives on how to maintain and enhance farm productivity using ecological techniques. But the ecological farmer also faces problems in preserving agricultural products and in bringing them to market. Two papers focus on this issue. Archileo Kaaya first describes the processing and storing of horticultural commodities in Uganda and makes some suggestions for improvements. Mbeza and Minde then look at some alternatives for the processing and storage of fruits and root crops in Malawi. These two papers just scratch the surface of a major problem that must be faced by all ecological producers as they seek to effectively commercialize their agricultural production.

The last section of the book turns attention to agricultural institutions. Yucabeth Ogega looks at bottlenecks to the dissemination of ecological agriculture in Kenya. Three papers then describe the work of training institutions in the region. Paul Desmarais describes the evolution of the training approach at the Kasisi Agricultural Training Centre in Lusaka, Zambia. Jailos Kanjanga describes the work of the Lipangwe Organic Manure Demonstration Farm in Malawi. And Livai Matarirano describes the work of the Fambidzanai Permaculture Centre in Harare, Zimbabwe. The book ends with Timothy Simalenga's review of the challenges faced by the Animal Traction Network for East and Southern Africa (ATNESA). Animal traction is an appropriate supplement to human labour on small farms and fits nicely into the mixed farming systems that provide the greatest potential for sustainable ecological farm management. ATNESA has been studying and promoting the use of animal traction since the early 1990s. Its experience offers important lessons to the group of farmers, researchers and activists seeking to mobilize an Ecological Agriculture Network for Eastern and Southern Africa.

The mixture of these ideas does not constitute a comprehensive guide to ecological farming. Rather the book offers a "sampling" of the wealth of options open to those who seek holistic, ecological answers. How can we define "ecological" or "holistic" in this context? What is the common thread that links these suggested paths and sets them apart from much of mainstream 20th century agriculture? Here we come to an issue where distinctions of geography, technological evolution and historical reference evaporate to reveal solid, permanent truths, truths which must be respected and assimilated by any society working toward an agriculture which will feed the people forever. We must work with nature, combining the wisdom of the past with all that modern science can elicit, we must create farming systems to compliment the unique circumstances of each locality, we must work toward local and regional self sufficiency, reducing dependency on outside inputs. We now have the benefit of painful evidence that short-term "shortcuts" to food production are not desirable or effective. The papers presented here reflect some of the variety of indigenous knowledge, current research and training efforts that will contribute to agricultural renewal in Eastern and Southern Africa as we look forward to the new millennium.

CHAPTER 1

ORGANIC MATTER, SOIL FERTILITY AND TRADITIONAL CROPPING SYSTEMS IN ZAMBIA

by Obed I.M. Lungu

INTRODUCTION

Fertile, healthy soil is characterized by its biological diversity and a perfect equilibrium between its biological, chemical and physical properties. This equilibrium is disturbed by human activity such as cultivation and removal of forest cover. The need for increased food production puts pressure on the soil ecology so that re-establishment of the equilibrium is either too slow or the quantities of plant nutrients released from natural cycles are not adequate to meet the demands of intensive cropping. Therefore, when land is limited, technologies which increase yields are adopted, which require heavy use of chemical fertilizers in order to supplement natural processes. However, these efforts to increase soil productivity are often not accompanied by an equal concern to maintain and enhance environmental quality. Some of the problems that arise include pollution of surface and ground water and the tainting of food chains by chemical fertilizers and pesticides.

In ecological soil management, the focus is on designing crop production systems that are well integrated and ecologically sound and that reduce or eliminate the need for chemical inputs. Ecology in an agricultural system refers to the cultivated land and its adjacent surroundings, the plants contained or grown thereon, soil microorganisms and all the animals associated with it – including humans (Gips, 1992). An ecologically sound agriculture is one whose overall health allows it to be resilient and productive over both the short and long terms, while ensuring a quality environment. There are many examples of traditional farming systems from Asia, Africa and Latin America that are ecologically sound, particularly in areas that have escaped foreign intrusion. Unfortunately, most of these systems have been dubbed "primitive" and are assumed to be unable to meet the demands for increased food production. The objectives of this paper are to describe some traditional cropping systems from Zambia and to discuss potential interventions for their improvement, while maintaining their ecological soundness.

The Role of Soil Organic Matter in the Maintenance of Soil Fertility

To sustain crop yields over time requires not only a plentiful supply of plant nutrients, but also the maintenance or even improvement of soil fertility. Fertile soils give consistently good yields without the addition of fertilizers and respond well to the application of fertilizers and other soil amendments if these are needed. Soil fertility is accumulated slowly through the complex interactions of the biological, chemical and physical properties of the soil. Soil organic matter is a key factor. Organic matter binds primary soil particles into stable aggregates, creating good soil tilth and reducing compaction, enhances plant nutrient and water retention and provides a substrate for microorganisms responsible for nutrient cycling through biomass. In fragile tropical soils, maintaining high levels of soil organic matter provides a key to sustaining soil productivity because these soils are generally acid with an inherently low capacity for nutrient and water retention.

Farming Systems in Zambia

There are three major farming systems in Zambia: traditional, emergent-permanent and commercial (Table 1.1). They are distinguished by the level of purchased inputs and the sophistication of the management practices used. In any of these systems, inept soil management practices that decrease soil organic matter content will result in soil degradation, leading to reduced soil productivity. Practices of removing crop residues from land after harvest or overgrazing will decrease soil organic matter content and will starve microorganisms of substrate, reducing their population and diversity. Bare soil not only exposes soil to erosion, but it also exposes soil microorganisms to dangerous sun radiation, or to predators. Monocropping, which is a common practice in commercial agriculture, gives microorganisms a monotonous diet. Under these conditions, the soil's natural fertility is lost.

Table 1.1 Major farming systems in Zambia and their characteristics

System	Fertilizer Input	Crops	Strategy for Improvement
1. Traditional			
a) Shifting	None	millets, sorghum, cassava, beans	Adapted varieties and species
b) Semi-permanent	0-Low	maize, beans, groundnuts	Lime, fertilizer, or low input technology
2. Emergent-permanent	Low-Medium	maize, soybeans, sunflower, cotton, groundnuts	Rotations and green manuring
3. Commercial plus Estates	Medium-High	Maize, wheat, soybeans, tobacco, citrus	Rotation and fertilizer use efficiency

Traditional cropping systems exemplify the greatest dependency on natural vegetation and natural cycles for the supply of plant nutrients for crop production (Lungu & Chinene, 1993). In these systems, all crop nutrients are released from organic matter. They thus exemplify elements of an ecological approach to soil management. These systems are invariably characterized by:

- a combination of high variety and species diversity
- structural and temporal diversity in the field
- efficient nutrient cycling and energy flows
- intricate complexities of biological interactions

Their specificity to climatic and ecological zones suggests recognition that not all soils are suitable for all crops. All the systems correctly aim to match crops with soil capability, thereby minimizing the requirement for external purchased inputs. These systems have evolved over centuries and this indigenous knowledge has been handed down from generation to generation, with relevant modifications as deemed prudent by the mentors.

Shifting cultivation

Axe/hand hoe system without livestock: Under shifting cultivation, land is cropped continuously for about six years before being left under bush fallow for up to 25 years, in order to allow the vegetation to regenerate and soil fertility to be restored. Millet, sorghum, beans, cassava and other crops are grown in various intercropping patterns. Minimum tillage with hand tools is practised. Without additional fertilizers soil fertility is exhausted within six years, hence the need to shift and clear virgin or secondary forest. Land supply is thus a critical limiting factor.

In agroecological zone III characterized by high rainfall (>1000 mm annually) and acid soils (pH <5.0), a peculiar variant called *Chitemene* has evolved. Land preparation involves lopping tree branches and gathering them in heaps to dry. They are then burnt releasing some nutrients from the biomass and the wood ash provides a beneficial liming effect (Table 1.2). Following light cultivation with a hoe, millet is planted in these patches. Other crops are grown for a further 2 years before cassava occupies the patches for two years until about the sixth year, when the land is again left under bush fallow for up to 25 years. Recropping of the land depends on the regeneration of soil fertility and arboreal regrowth.

Table 1.2 The long-term effect of woodburn on soil properties under "Chitemene" at Misamfu Regional Research Station, Zambia

	Depth cm	pH	Organic C %	C/N Ratio	Available P mg kg^{-1}	Exchangeable Bases cmol kg^{-1}		
						Ca	Mg	K
Undisturbed before burning	0-10	5.2	2.07	14	12.0	1.38	1.67	0.25
	10-20	4.7	0.82	9	2.0	0.86	0.37	0.17
Ash		10.5			28.0	11.51		22.0
39 hr after burning	0-10	6.2	1.71	5	87.0	7.15	3.62	0.88
	10-20	6.3	0.94	10	45	1.95	2.96	0.21
3 1/2 years after burning	0-10	4.8	1.40	20	14	3.45	0.70	0.15
	10-20	4.6	0.75	10	11	3.0	0.55	0.10

Source: Stroomgard, 1991.

This is a very extensive system, with the ratio of cleared land to cultivated area ranging from 5 to 20%, and the cropped land takes a very long time to recover under fallow. The heat destroys soil organic matter and leads to loss of nitrogen. The heat also sterilizes soil, reducing microbial activity. Irreversible dehydration of soil colloids may occur through destruction of organic coatings on soil particles. Thus soil structure is destroyed, and the soil loses capacity to retain nutrients and water. Stroomgard (1991) has shown that 4 years after the burn the soil is again acid and much depleted in plant nutrients (Table 1.2).

Axe/hand hoe with livestock present: In agroecological region IIb (800-1000 mm annual rainfall), there is low nutrient and water retention of the sandy soils derived from the Kalahari sand parent material. A viable traditional system that has evolved there uses cattle. The cattle are kept in night kraals in open fields and are relocated several times a year. Cropping is carried out on the patches where dung has been mixed with soil, creating a zone of concentrated plant nutrients. These areas are very productive for one or two seasons and then they are abandoned when soil fertility declines, partly due to loss of nutrients through leaching.

Semi-permanent farming systems

An alternative way to release plant nutrients from biomass entails composting vegetation in mounds or ridges. Ridging is a common practice in all hand-hoe farming systems. Crop residues are usually buried in mounds after harvest during the dry season, and crops are planted on the ridges during the following rainy season. Alternatively crops are planted on flat land, but at weeding the crops are ridged and weeds are buried in the ridges. Soil depth is increased by the ridging, and in very wet years this practice can save crops from excessive moisture. Clearing land from virgin forest involves ring-barking trees; the stumps are left intact in the fields. Some tree stumps do regenerate annually and the arboreal regrowth is added to the weeds in the mounds or ridges. In one variation of the practice, sweet potato is the first crop on the mound or ridge. In order to obtain sufficient biomass, the operation is commonly carried out after good rains, when there is abundant fresh vegetation. Subsequently, the mounds and ridges are leveled and other crops are planted on the flat.

Hand-hoe rotational grass mound system (*Fundikila*): In the *Fundikila* system, natural vegetation, including nutritionally-poor *Hyperrhenia spp*, is buried in single circular mounds or in ridges during the dry season. Beans are planted on the mounds or ridges during the following rainy season. In the second year, the mounds are leveled and millet is planted on the flat. This alternate cropping on mound and flat continues for several years, with little or no application of chemical fertilizers. Stroomgard (1991) monitored some changes in soil chemical properties occurring under this system, and the data are reported in Table 1.3. Compared to the *Chitemene* system, the changes in soil chemical properties under the *Fundikila* are nominal, especially with regard to pH, available P and exchangeable bases (Ca, Mg and K). Over a period of 4 years of continuous cropping, these properties showed smaller seasonal fluctuations than under *Chitemene*.

Table 1.3 The long-term nutrient accumulation under natural grassmound composting (Fundikila) at Misamfu Regional Research Station, Zambia

	Depth cm	pH	Organic C %	C/N Ratio	Available P mg kg^{-1}	Exchangeable Bases cmol kg^{-1}		
						Ca	Mg	K
Undisturbed fallow	0-10	5.2	1.0	25	2.0	1.2	1.4	0.2
	10-20	5.1	0.6	2	1.0	0.7	1.1	0.3
2 months mounds	0-10	4.9	0.9	23	4.0	1.2	1.0	0.3
	10-20	4.9	0.9	15	2.0	1.2	0.9	0.2
2 years flat	0-10	5.0	0.7	18	2.0	1.1	0.8	0.1
4 years flat	0-10	5.3	1.1	-	-	2.4	1.1	0.2
	10-20	5.0	1.1	-	-	1.8	1.0	0.3
	10-20	4.8	0.5	17	3.0	0.6	0.6	0.1

Source: Stroomgard, 1991.

Hand-hoe and ox-plough: This group of farming systems is practised in about one quarter of the cropped land in Zambia, and it is dominant in agroecological zone II. Livestock play an important part in the interaction with cropping and forestry. Fresh vegetation and crop residues are buried in soil by inversion with an ox-drawn mouldboard plough. Because of the limited draft power from oxen, ploughing is usually delayed until the soil is sufficiently wetted by rain, at which time the natural vegetation would be abundant. The burying of the organic matter by inverting the soil produces the same effect on soil chemical properties as grass mounding or ridging.

Potential Interventions to Improve Soil Fertility in Traditional Systems

One major criticism of these traditional cropping systems is that they are less productive than systems in which there is heavy application of chemical fertilizers and use of other condiments of modern agriculture. For example, yields of hybrid maize may rarely reach 2 t/ha using traditional methods compared to 7-8 t/ha with fertilization under Zambian conditions in a normal year. Unimproved indigenous varieties of crops that are widely planted in traditional cropping systems yield worse still at less than 1 t/ha. There are several aspects of traditional cropping systems in which intervention with new technologies could enhance soil productivity and ensure sustainable crop production.

Improving nutrient supply

Nutrient supply directly affects crop production. Natural soil fertility will give a base yield, but any increase in yield above base yield will depend on continuous management of the supply of adequate nutrients. Addition of plant nutrients through incorporation of biomass is the basis of traditional systems and can be improved.

- **Green manuring:** Replacing natural bush/grass fallows with nitrogen fixing plants will increase the supply of N to the soil. This practice has the potential to shorten the fallow period to less than the current 25 years in shifting cultivation systems. A three-year fallow with the shrub *Sesbania sesban* has been shown to support two consecutive crops of maize without application of chemical fertilizer. The yields of maize following such a fallow in Eastern Zambia were 7-8 t/ha compared to 3-3.6 t/ha with application of 110 kg N, 40 kg P_2O_5, and 20 kg K_2O per ha (Kwesiga & Cole, 1994). In field trials at Makoka Research Station in Malawi, another shrub, *Tephrosia vagelii*, has also been shown to sustain maize yields of 3-5 t/ha for two consecutive seasons on the same plot previously planted to the shrub (Maghembe, Chirwa & Kooi, 1995). According to Mwambazi (1994), bean yields on an acid soil in Northern Zambia more than doubled (1000 to 2300 kg/ha) when the crop followed a green manure crop of sunhemp (*Crotolaria zanzibarica L.*). The benefits of these shrubs include recycling nutrients from the subsoil and biological nitrogen fixation. With *Sesbania sesban*, nitrogen fixation rates of up to 350 kg/ha/yr have been recorded in India (ICRAF/NFTA 1989). Green manuring also increases the availability and supply of other nutrients. Incorporating 8 t/ha of sunhemp residues into an acid soil resulted in significant increases in soil pH and levels of Ca, Mg, and K (Lungu & Dynoodt, 1997). The Ca, Mg and K can be efficiently recycled through biomass by planting legume species with varying rooting depths in rotation with shallow-rooted food crops.

Farm yard manure: Application of large amounts of chemical fertilizers to improve soil fertility can hardly ever increase yields to levels obtained on soils with many years of farm yard manure application. Data from a long-term field trial at Rothamstead Experimental Station, in which the effects of chemical fertilizer and farm yard manure on the yield of a high-yielding variety were compared, showed greater yields on plots with manure than on those receiving only chemical fertilizer, however much nitrogen the latter received. The data suggest that improved soil fertility provided by several years of manure application contributed to soil productivity. One other benefit is the amelioration of soil acidity. Lungu *et. al.* (1993) showed that incorporation of 30 t/ha of farm yard manure into an acid soil gave the greatest increase in plant growth and yield, compared to lime alone at the recommended rate of 6 t/ha. Because traditional cropping systems conserve soil organic matter, soil acidity has rarely been a major problem, especially with a careful selection of acid tolerant crops such as cassava, millet and sorghum.

Several areas of nutrient management need further research in order to improve the efficiency of plant nutrient supply from biomass. One aspect concerns synchronizing nutrient release from organic matter to crop growth periods of maximum nutrient demand. The effect of some residues in soil is short-lived, for example only two months under field conditions for sunhemp (Lungu & Dynoodt, 1997). Besides, breeders have increased crop yields by improving earliness in high-yielding varieties, which may affect the efficiency by which modern cultivars use N from green manures and organic wastes. Different optimum plant populations may also have to be determined for use with these nutrient sources.

Maintaining soil organic matter and soil cover

The organic matter content can be improved through a regular application of farm yard manure, through the practice of minimum tillage and through rotational cropping that includes plants with varying rooting depth. Organic matter decomposes at a constant rate, and in the tropics this is so rapid that net organic matter accumulation is difficult to achieve in a short time. However, by regularly applying farm yard manure and other organic materials, the levels of soil organic matter can be maintained and progressively increased (Lungu, 1987).

- **Conservation tillage:** The process of building up soil organic matter can be enhanced by employing conservation tillage techniques. Use of a mouldboard plough has several advantages over the disc plough. A mouldboard plough, especially ox-drawn, which is normally used under moist soil conditions, inverts sod and any organic residues. In contrast, a disc plough behind a tractor and used under dry soil conditions, pulverizes the soil rendering it vulnerable to water and wind erosion and to crusting and compaction after rain or irrigation.

- **Mulching:** Unlike traditional cropping systems, most conventional tillage systems usually make no attempt to preserve soil cover. Crop residues are burnt or removed and fed to livestock. Water erosion from bare arable land results in an annual soil loss estimated at 3 t/ha in Africa, which impoverishes topsoil by the loss of nutrients. Soil cover can be restored through using mulches in combination with minimum tillage practices.

Retaining crop residues (stover) on the land after the harvest has been shown to maintain soil productivity under continuous cropping. In a six-year continuous cropping trial in Zambia, it was shown that maize yields with application of chemical fertilizer could be maintained by leaving crop residues on the land (Figure 1.1). Minimum or zero tillage and agroforestry alley cropping practices, in which

Figure 1.1 Yield of maize in continuous cropping with and without return of stover

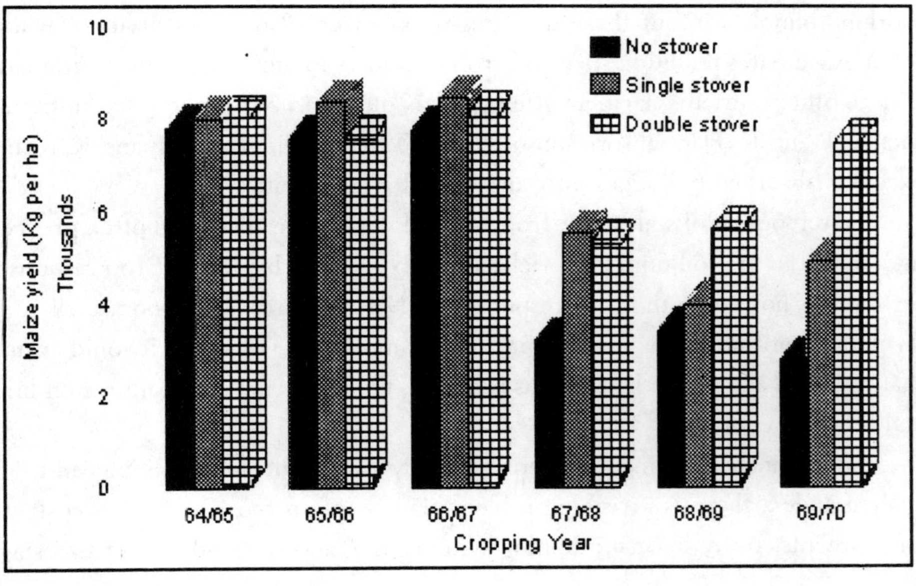

plant residues are retained or imported onto the land, have all proven to maintain or even enhance soil productivity. With minimum tillage, considerable amounts of root biomass remain in the soil if it is not disturbed by excessive land preparation as occurs under conventional cultivation practices.

However, mixed cropping, retaining crop residues, or recycling organic wastes onto land are not popular practices with farmers. The commonly cited problems are that residues are not compatible with mechanization at the farm due to different planting and maturity dates, as well as the potential spread of disease, pests and weeds. In order to popularize this aspect of ecological soil management practice, some economically viable solutions must be found. The effectiveness of these solutions will have to be practically demonstrated under farm conditions.

Monitoring Soil Fertility

Monitoring soil fertility is essential to the implementation of ecological soil management practices. Regular measurement of soil pH, soil organic matter and available P, Ca, Mg and K can provide the information needed to evaluate the fertility status of the soil (Lungu, 1987). Monitoring can provide important information on nutrient balances and soil dynamics. This information is useful in planning cropping, setting realistic target yields and in guiding fertilization programmes so that remedial action can be taken before the soil is irreversibly degraded. Fertilizer use can be reduced while maintaining soil productivity. Working blindly without these data can result in unbalanced fertilization, which often exacerbates pollution. Because crop responses to nitrogen are more dramatic than to other nutrients, farmers often apply N out of proportion to other nutrients such as P and K. Plant absorption of N would then be limited by P and K, so the excess unabsorbed N leaches into surface and ground water.

Development of soil acidity from applied nitrogen fertilizer can often proceed undetected in the soil until crop yields drop to unprofitable levels. Crop responses to lime do not equal those of responses to N so that resource-poor small-scale farmers do not normally purchase it. Regular monitoring of soil pH would ensure that remedial action can be taken as required, thus reducing expenditures on lime and other ameliorants.

Unfortunately, farmers do not regularly carry out soil sampling and soil analysis unless there is a visible problem. This is often too late to save costs on amendments. In traditional cropping systems, the land is abandoned at this stage and left for long periods to regenerate under bush fallow. Furthermore, most farming activities commence without first conducting soil surveys and land evaluation so as to match cropping activities with soil capability. The potential for soil degradation is greatest under these situations. Extension workers and researchers can make an important contribution to farm productivity by supporting soil monitoring by farmers.

Crop diversification

Diversity is a natural phenomenon and ecological soil management systems should strive to maintain and enhance this diversity. Crop diversification has several advantages. It is a form of food security. Growing different varieties of the same crop or even different species in the same field lessens the danger of crop failure due to adverse weather and other environmental factors such as disease (Ferguson & Mkandawire, 1996).

Crop diversity combined with properly managed crop sequencing in fields also maintains and improves soil fertility. Maintaining crop variety and species diversity on the same land through spatial and temporal arrangements can ensure protective soil cover throughout the year. Rotation of deep and shallow-rooted plants increases soil depth and reduces compaction, while helping to recycle nutrients. Deep soil, rich in organic matter, is desirable because this increases water and nutrient retention and absorption, which can be critical to obtaining good yields in drought conditions. Above ground, a canopy comprising tall and short crop varieties or species with different maturity dates ensures soil cover throughout the year and also optimizes on the interception of energy from the sun.

Use of different varieties of the same crop or species on the same land either in space (e.g. mixed cropping) or in time (rotation) optimizes plant nutrient use efficiency. Different plants exhibit different physiological responses to environmental stress such as nutrient deficiency. For instance, Lungu (1983) showed that buckwheat (*Fagopyrum esculentum L.*) grown in P-deficient soil or in a P-deficient nutrient solution, acidified these media and released organic root exudates. Consequently, insoluble soil nutrient sources, tested by ground rock phosphate in the experiment, became soluble and available for plant uptake.

These findings have potential use for cropping systems where annual application of chemical fertilizers have led to a substantial accumulation of unavailable nutrient forms, especially of P and K. At Rothamstead, it was shown that with certain crops, application of large amounts of P and K to improve yields on impoverished soils could hardly ever increase yields to levels obtained on soils containing several years of residual P and K.

Conclusion

Failure to maintain adequate levels of soil organic matter in the soil, especially in nutrient-poor tropical soils, can initiate and perpetuate soil degradation, leading to reduced soil productivity. But maintaining high levels of organic matter in soil is not easy, especially under tropical conditions favourable to rapid decomposition of organic matter. There are many examples of farming systems from Africa that have traditionally optimized soil organic matter content for crop production. Traditional cropping systems should be considered as providing an opportunity for creative integration of local empirical knowledge systems with disciplines of scientific research. Herein lies the challenge to agricultural researchers and extension workers: How to design soil management systems that increase crop yields, while maintaining and enhancing environmental quality and ensuring sustainable production over both the short and long terms. The potential of residue management, improved fallow with green manures, grass compost mound systems, soil monitoring and crop diversification have been presented. Rural small holders are confident working with familiar techniques and with materials that are available locally. Continuing research with farmers can improve on these traditional techniques in ways adapted to local conditions. Researchers can support significant sustainable improvements in productivity by concentrating on the needs and the existing knowledge base of the resource-poor smallholder.

Building and maintaining high levels of soil organic matter with these ecological management techniques represents a capital investment by the farmer. An investment which will be paid back over a period of years. Security of tenure on the land is thus a key factor, providing an incentive for farmers to practise ecological soil management. When farmers own the land, they are more likely to adopt techniques that will allow them to farm the land in a more sustainable way.

References

Ferguson, A.E. and Mkandawire, R.M. 1996: A crop diversity improvement strategy. *Indigenous Knowledge and Development Monitor* Vol. 4, Issue I (April), 6-7.

Gips, T. 1992: What is sustainable agriculture? Part One: Ecological soundness.

MANNIA Spring, 1992, 5.

ICRAF/NFTA 1989: Executive summary of the proceedings of the International Workshop on Perennial Sesbania Species in Agroforestry Systems. Nairobi: ICRAF.

Kwesiga, F. and Cole, R. 1994: The effect of short rotation *Sesbania sesban* planted fallows on maize yield. *Forest Ecol. & Manage*. 64:199-208.

Lungu, O.I.M. 1983: Differences in the phosphate uptake mechanisms of P-efficient buckwheat (*Fagopyrum esculentum L.*) and responsive wheat. Unpublished Doctor of Philosophy Thesis. Davis: University of California.

Lungu, O.I.M. 1987: Monitoring soil fertility. In *AFRICALAND-Land Development and Management of Acid Soils in Africa II*, Proceedings of the second regional workshop held at Lusaka and Kasama, Zambia from 9-16 April 1987, pp. 161-172. IBSRAM Proceedings No. 7.

Lungu, O.I. and Chinene, V.R.N. 1993: *Cropping and soil management systems and their effect on soil productivity in Zambia: a review*. Ecology and Development Paper No. 5. Aas, Norway: Centre for Sustainable Development, Agricultural University of Norway.

Lungu, O.I., Temba, J., Chirwa, B. and Lungu, C. 1993: Effects of lime and farm yard manure on soil acidity and maize growth on an acid Alfisol from Zambia. *Tropical Agric*. (Trin.) 70, 309-314.

Lungu, O.I. and Dynoodt, R.F. 1997: *Effect of green manure and chemical fertilizer on nutrient dynamics and yield of maize with continuous cropping*. Unpublis manuscript.

Maghembe, J.A., Chirwa, P.W. and Kooi. G. 1995: *Annual Report for 1995 SADC/ICRAF Agroforestry Project, Zomba, Malawi*. Nairobi: ICRAF.

Mwamabazi, T.N. 1994: Utilization of organic plant materials for soil improvement in the high rainfall areas of Zambia. Unpublished. Misamfu Regional Research Station, Soil Productivity Research Programme.

Stroomgard, P. 1991: Soil nutrient accumulation under traditional African agriculture in Miombo woodland of Zambia. *Tropical Agric*. (Trin.) 6, (1) 74-80.

CHAPTER 2

THE *NGORO* CULTIVATION SYSTEM: SOIL PRODUCTIVITY ON STEEP SLOPES IN SOUTHWESTERN TANZANIA

by Zacharia J.U. Malley[1]

INTRODUCTION

Ngoro is a local name for the pitting practice of the Matengo tribe in the southwestern highlands of Tanzania, popularly known as the Matengo highlands. The *ngoro* system is also referred to as the pitting system, Matengo pits or the pit cultivation system of Wamatengo. The system is an indigenous way of preparing land for cultivation on steep slopes. This cultivation system consists of pits surrounded by ridges, with crops grown on the ridges around the pits.

The men begin the work of creating the *ngoro* by slashing grass and other plant residues. The slashed materials are allowed to dry for about 14 days and are then arranged in rows along and across the slope. Between the rows are one- or two-metre squares of soil, forming a matrix over the whole land area. The women use hand hoes to dig the soil in the centres of the squares and throw it on the rows of dry grass to form the ridges. Thus the ridges consist of a sandwich of undisturbed soil below, a layer of dried grasses and raised soil on top. Pits are formed in the centre of the squares. The system is usually constructed late in the rainy season in March/April, when the soils are very wet and the rains are about to end.

This paper reports on some of the *ngoro* cultivation system's ecologically sustainable features for improving agricultural production. The pros and cons of the system and possible applications of some of its features in other tropical environments are discussed. Further areas of research interventions are suggested.

BACKGROUND

The *ngoro* cultivation system is unique to the Matengo highlands in Mbinga District

(Ellis-Jones, Kayombo, Martin, Dihenga & Thadei, 1994). The Matengo highlands are characterized by steep hills and narrow valleys (Stenhouse, 1944; Allan, 1965; ICRA, 1991). The main vegetation of these highlands are grass and a few surviving remnants of the former miombo forest found in small patches on the mountain tops. The area's climate is temperate tropical with a unimodal rainfall pattern (Ellis-Jones *et al.*, 1994). Rains start in November and stop in May of the following year. The rest of the year is dry. Total annual rainfall ranges from less than 1000mm to over 1600mm, with an average of 1224mm. The average minimum and maximum temperatures are 19-23C and 29-31C, respectively, depending on the micro-climates within the topographical variations.

The population density in the mountainous areas has reached 120 people/km2, which is the highest in Mbinga District (Ellis-Jones *et al.*, 1994). In order to cope with the increasing food requirements of the growing population, more land on the hill and mountain sides is being cleared and put into cultivation. This accentuates familiar soil degradation processes found in hilly and mountainous highlands. Degradation is further accelerated by the type of soil. The soils of the area are mainly haplic or humic acrisols (oxisols and ultisols), depending upon their position on the land topography (UAC, 1989). Under vegetation cover, acrisols (oxisols and ultisols) have low bulk densities and thus a favourable soil structure. This is closely related to flocculation and cementation of clay by organo-mineral complexes to form stable aggregates (Lal, 1986). However, this soil structure breaks down easily once the vegetation is removed. The removal of vegetative cover results in several problems:

- soil organic matter losses
- severe compaction due to collapse of the soil structure and acquisition of a hard consistency upon drying
- shallow effective rooting depth for seasonal crops
- severe chemical and nutritional problems caused by nutrient depletion from runoff and leaching
- soil erosion through runoff
- poor moisture conservation capacity
- soil acidification due to loss of organic matter with washing of basic cations
- increased soil temperature as the soil is exposed to direct sun

These degradation processes can lead to permanent and irreversible loss of soil productivity (Dregne, 1990). Denuded hill and mountain slopes store less water than slopes covered by thick humus and spongy soils. As a result, streams can dry up and then become destructive torrents in the rainy season due to huge volumes of runoff from hill and mountain slopes. Unprotected denuded sloping land sheds runoff and debris down the slope, depositing them on otherwise fertile lowland (Lowdermilk, 1986). Silt in sediment chokes streams, resulting in overflows which can cause loss of life and property. In newly settled areas of the Matengo highlands, where less *ngoro* cultivation is practised, these processes are degrading the soil and water. But the *ngoro* system used on the steep slopes of the Matengo highlands has the potential to circumvent most of these problems.

HISTORICAL EVOLUTION OF THE *NGORO* SYSTEM

The *ngoro* system was developed by the Matengo tribe who live in the hilly areas of Mbinga District. Wamatengo were initially driven to these inaccessible hilly areas by the raiding Ngoni tribe of Zulu-Swazi origin and by slave traders (Stenhouse, 1944; Allan, 1965). The Wamatengo settled near large natural caves on inaccessible mountain tops where they could defend themselves from these raiders. There the Wamatengo produced food crops for themselves as well as food tributes for Wangoni raiders. The Wangoni were raiding the Matengo people to steal foodstuffs and women. The women were protected in the caves and guarded by their husbands when they worked in the field. This may account for the traditional division of labour along gender lines in the construction of pits, as well as the traditional belief that good *ngoro* can only be made by true Matengo women.

Near the caves at Langiro, the first defensive mountain position occupied by the Wamatengo, Stenhouse (1944) reported badly eroded hill slopes with no trace of pit cultivation. He also reported the presence of exposed large tree roots. These facts suggest that when the Matengo people initially settled at Langiro, they cleared forests and cultivated the steep slopes, using the conventional system for cultivation on flat land. This resulted in large volumes of runoff and soil erosion. This is the likely cause for developing the pit system, because the second defensive position taken by Wamatengo, near the caves at Litembo, showed no sign of soil loss by erosion for over sixty years under cultivation using the pit system. This area is said to be the origin of the pitting technique. Around 1890, the Germans found large concentrations of Matengo people in the fertile Litembo area, all practising the *ngoro* system (Stenhouse, 1944; Allan, 1965). According to Allan (1965), Wamatengo developed this highly effective agricultural system where there was a shortage of suitable land for cultivation. They had only two options: cultivate the steepest slopes or starve.

IMPORTANT ECOLOGICAL FEATURES OF THE *Ngoro* SYSTEM

- **Soil surface roughness:** Rough soil surfaces are known to be effective for controlling runoff and soil erosion (Romkens & Wang, 1986), improving infiltration (Larson, 1962; Zobeck & Onstad, 1987) and decreasing the evaporation rate (Linden, 1982). Generally, soil surface roughness is closely related to surface depression storage capacity (Monteith, 1974; Onstad, 1984). In the *ngoro* system, pits create depression roughness and soil losses through surface runoff are reduced by in situ trapping of detached soil particles in pit bottoms.
- **Prevention of soil losses:** Preliminary results of a study in Mbinga District by Thadei (1995) showed that the *ngoro* system was superior to ridges for reducing the sediment loads collected from 9% and 20% slopes. Mututho (1989) showed that a similar pitting practice in Kenya was superior to conventional terracing in harvesting runoff and trapping silt for eventual maize cropping on denuded grazing land. Gichangi, Jones, Njarui, Simpson, Mututho & Kitheka (1989) reported that Katumani pitting (a modified *ngoro*) was extremely effective in rehabilitating eroded land. They found that the Katumani pits effected this through drastic reduction of runoff, enhanced soil water status in the pits near the trenches. This enabled the production of substantial yield of cowpea grain and various forage legumes in good seasons. Even in seasons of poor rainfall, the elimination of runoff resulted in enhanced total dry matter production, although the grain yield was not significantly higher than the control treatment. Comparable results were reported by Allan (1965) for northern Tanzania.
- **Conservation of surface runoff into the soil:** High rain runoff on the steep slopes reduces the soil's productivity. Conserving rainfall as soil water on steep slopes is an important aspect of soil management. When the rain can infiltrate and be held in the soil, it increases the amount of water available for crop growth and prevents soil degradation by erosion (Lal, 1986).
  The *ngoro* system can control runoff and soil loss by up to 60% (Ellis-Jones *et al.*, 1994). During the 1995/6 rainy season, Malley used three different pit sizes to demonstrate that, regardless of pit size, the *ngoro* cultivation system enhances soil water status. Soil surveys by Msanya, Kimaro & Magoggo,

(1995); Kimaro, Meliyo, Msanya, Magoggo, & Wickama (1996) and Msanya, Wickama, Kimaro, Magoggo & Meliyo (1996) indicated that the Matengo highlands soils have an available water capacity of between 120 to 200 mm/metre of soil profile. Malley found that available soil moisture capacity under *ngoro* practice exceeded those figures. The *ngoro* pits showed levels ranging between 245 and 286 mm/metre, with means between 264 and 267 for *ngoro* pits of various sizes from one to four metres square (Malley, 1997). The *ngoro* system of cultivation seems to improve water retention by reducing both runoff during the rains and evaporation after the rainfall. The pit system improves infiltration by catching the runoff, reducing water and soil losses. Monteith's studies (1974) showed that a high surface depression retains water in excess of infiltration from short duration storms. Additionally, pit roughness enhances water retention by reducing the evaporation of the water stored in the pit depressions. This is due to the creation of a sheltered microclimate that results in a stable air/water interface, which reduces water evaporation in the pits. The *ngoro* system improves environmental quality in two ways. It traps runoff water which recharges subsoil water for plant growth. The trapped runoff drains into the soil and feeds the underground water pool, providing a constant supply of water to streams, which are an important source of water in the Matengo highlands.

- **Soil organic matter, bulk density and fertility:** The incorporation of organic materials in the ridges is another important feature of the *ngoro* cultivation system. The plant materials are slashed and arranged in rows, running along and across the slope, forming squares of soil. Topsoil slices are cut from the soil squares and thrown upside down on the rows of dried grasses. Soil rich in organic matter and low in bulk density is concentrated in a 5 to 15cm soil layer on the *ngoro* ridges. This is then covered by subsoil with less organic matter and high bulk density excavated from the centre of squares. The digging creates the square pits.

This distribution of the soil organic matter and bulk density in the *ngoro* ridges appears to help preserve soil productivity in several ways:

- The productive topsoil is concentrated in the ridges on which the crops are grown, giving the seedlings nutrients, moisture and aeration.
- The ridges are covered with less fertile subsoil. This cover protects fertile soils and reduces soil organic matter losses due to erosion.

- The low bulk density in the 5 to 15 cm soil layer of the *ngoro* ridges improves porosity of this layer. This enhances rain water infiltration into the ridges reducing runoff. Good soil organic matter concentration in the 5 to 15 cm soil layer of the ridges improves soil water retention in this layer.
- The grass layer under the top soil reduces surface water flow by allowing runoff seepage to occur before the pit overflows.

Malley's analysis of the association between soil organic matter and soil bulk density on *ngoro* ridges showed that 86.2% of the variations in the bulk density could be described by variations in the soil organic matter concentration (Table 2.1)[2]. The inverse relationship between organic matter and bulk density implies that the soil porosity in the *ngoro* ridges is improved by organic matter.

Table 2.1 Soil organic matter and bulk density in 0-24 cm soil layers of *ngoro* ridges

Depths (cm)	N		Organic matter (%)	Bulk density (Mg/m3)
0 - 5	36		3.67±0.3	1.11±0.03
5 - 10	36		4.29±0.3	1.04±0.03
10 - 15	36		4.28±0.3	1.04±0.03
15 - 20	36		3.31±0.3	1.13±0.03
20 - 25	36		2.33±0.3	1.22±0.03
0 - 30		(initial)	2.96	1.30

Soil organic matter content is also an index of potential fertility of the soil. Therefore, the results in Table 2.1 suggest that the *ngoro* system of cultivation improves and preserves soil fertility. This is supported by the results of Martin, Willcocks & Kayombo (1996) showing that total nitrogen was highest in 5-10 cm soil layer in the *ngoro* ridges.

- **Penetrometer resistance:** The *ngoro* ridges consist of an undisturbed soil layer under a grass layer with a mechanically loose soil layer on top of the grass layer. The mechanical loosening of the soil is one of the important ways of improving a crop's ability to set roots (Kayombo, 1992). Penetrometers are used to provide a relative measure of the resistance offered by soil to the penetration of roots. Different types of penetrometers have been widely used to investigate soil physical properties and their effect on plant growth (Barley & Greacen, 1967). The field cone penetrometer test is the most common and best means used to detect *in situ* differences in rootability or root impeding zones of soils (Bengough, 1991). A soil with penetrometer resistance of 1 MPa is sufficient to cause a substantial reduction in root growth rate, and values greater than 3 MPa can halt or impede root growth greatly (Bengough & Mullins, 1990). The penetrometer resistance tends to be lower in soils with low bulk density and high porosity (Bauder, Randall & Swam, 1981; Bowen, 1981; Meek, Rechel, Carter & DeTer 1992). High soil bulk density and low porosity reduces rates of root growth and development (Barley & Greacen, 1967).

A study of penetrometer resistance on the *ngoro* ridges showed that the soil on top had lower penetrometer resistance than the uncultivated subsoil (Table 2.2).

Table 2.2 Mean penetrometer resistance values (MPa) on a 15% slope in different *ngoro* sizes at different depths (means of 68 observations)

	Pit sizes		
Soil layers (cm.)	1m. x 1m.	1.5m. x 1.5m.	2m. x 2m.
0.0 - 7.5	0.60	0.58	0.35
7.5 - 15.0	0.83	0.78	0.69
15.0 - 22.5	1.06	1.01	0.87
22.5 - 30.0	1.27	1.27	1.12
30.0 - 37.5	1.40	1.39	1.27

This reduction in penetrometer resistance is closely and negatively associated with maize height 35 days after planting and with maize yield (Table 2.3). This shows that the *ngoro* system alleviates the soil compaction problem.

Table 2.3 Linear regression equations showing relationships between soil penetrometer resistance with plant height and grain yield on a 15% slope in the Matengo highlands.

Soil Layer (cm)	Linear Regression Equations	r2
	Plant heights (m)	
0.0 to 7.50	PH = 0.402 - 0.285PR	0.985
7.5 to 15.0	PH = 0.713 - 0.594PR	0.995*
15.0 to 22.5	PH = 0.671 - 0.423PR	0.971
22.5 to 30.0	PH = 0.798 - 0.444PR	0.827
30.0 to 37.5	PH = 0.997 - 0.547PR	0.876
	Grain yield (t/ha)	
0.0 to 7.50	GY = 1.866 - 0.315PR	0.961
7.5 to 15.0	GY = 2.180 - 0.628PR	0.992
15.0 to 22.5	GY = 2.144 - 0.454PR	0.999*
22.5 to 30.0	GY = 2.306 - 0.497PR	0.928
30.0 to 37.5	GY = 2.518 - 0.605PR	0.960

r2 = coefficients of determinations; PR = penetration resistance in MPa;
PH = plant heights; GY = maize grain yield; * = significant at p = 0.05.

Crop Rotation, Fallowing and Mulching

Crop rotation and bush-grass fallowing practices are important soil management techniques for improving soil productivity. These have a variety of effects on the soil:

- Improved soil tilth.
- Increased soil organic matter.
- An improved supply of nitrogen to the soil by legume plants through fixation of atmospheric nitrogen in their root systems.
- Higher nutrient absorption from different soil zones by different plants during the rotation and fallowing period. This is important because deep rooted crops/plants absorb certain nutrients from the subsoil and return them to surface soil through their leaf fall and root residues. The returned nutrients may then be available for shallow rooted crops.
- Prevention of build-up of weeds, pests and soil-borne diseases that are specific to a given cropping system.

The *ngoro* system has a characteristic crop rotation and fallowing practice. The International Centre for Development Oriented Research in Agriculture (ICRA) (1991) reported that the main rotation associated with the *ngoro* system is fallow-beans-maize-fallow-beans. Maize is sown in planting holes on the *ngoro* ridges in December, following a July-to-November fallow and after the bean crop planted in March has been harvested. Beans are sown by broadcasting on newly-half formed *ngoro* ridges in March, after fallow following the harvesting of maize in August. Broadcasted beans are covered with subsoil from the pit during the process of pit construction. However, this rotation system may vary from one place to another, with beans being replaced with other crops such as cassava and wheat. Fallow periods are short in areas with high population densities.

In the *ngoro* cultivation practice, maize stubble is left on the soil surface after maize harvesting in August and the stubble remains on the surface throughout the fallow period (September to February). The stubble protects the soil from erosive rains and helps reduce the speed of surface runoff. This results in increased soil water available for the following bean crop which is sown on the semi-formed *ngoro* ridges towards the end of the rainy season in March. Decomposition products of the stubble also improves soil fertility and physical conditions (Unger, Stewart, Parr & Singh, 1991).

Stewart and Robinson (1996) observed that the loss of organic matter is perhaps the single most degrading factor to soil quality in hot and dry areas.

They further observed that organic matter is lost faster in areas with high temperatures, and, because precipitation is limited, the amount of organic matter returned to the soils is low. This situation is addressed in the *ngoro* system through *in situ* composting of organic residue, crop rotation, fallowing and stubble mulching. Where conditions permit, these techniques could be applied in other environments for maintenance and/or improvement of soil productivity.

Conclusions

The *ngoro* cultivation system has preserved the soil productivity on steep slopes in southwestern Tanzania. The most important features of the system are:

- high surface roughness created on the steep slopes for trapping runoff and soil particles;
- *in situ* composting of organic residues during pit construction improving soil organic matter;
- use of crop rotations, fallows, and stubble mulching to increase soil fertility, water absorption and runoff infiltration;
- the formation of bunds around the pits during pit construction, which loosens the naturally compacted soils and allows for easy rootability.

In considering the advantages and disadvantages of the *ngoro* system, Ellis-Jones *et al*. (1996) recognized the system's numerous advantages.

- It helps to maintain soil fertility for longer periods without fertilizer applications.
- It effectively controls surface runoff, reducing soil erosion.
- It increases moisture availability for crop use towards the end of the rainy season.
- It is re-constructed only every two years, saving some labour for one year.
- It creates a small micro-climate within the pits which may enhance crop production.
- It reduces weeding requirements as weed seeds are buried deep in the *ngoro* ridges.
- It provides the greatest returns as a low-input system.

But they also recognized that the system had some disadvantages.

- It needs long fallow periods to maintain soil fertility, but the long fallows cannot currently be used due to increased population pressure on the land.

- The incorporated organic material in its ridges causes crop microbes to compete for inherent soil N before nutrients are released from the organic matter.
- High labour is required for pit construction every second year.
- As the system consists of pits and ridges, the planting area is reduced, resulting into low plant population per unit area.
- It is difficult to mechanize the construction of the *ngoro* system.

Due to increasing population pressure on the land, the fallow period is declining and less organic residue is being incorporated, resulting in lower productivity. This suggests the need to test additional strategies for improving productivity. The use of legume-enriched short fallows using herbaceous/shrub plants such as *Crotalaria spp* and *Sesbania sesban* as part of the rotation should be explored. The possibility of using green manure plants during short fallows should be considered. Finally, a study on the potential for the use of supplemental fertilizers should be undertaken.

Endnotes

1. The author wishes to thank the Principal Secretary for the Ministry of Agriculture, Tanzania; Zonal Director and Zonal Research Coordinator for the Southern Highlands Zone for permission to participate in the Roundtable on Ecological Agriculture. The author would also like to thank the head of the maize improvement programme at Marti-Uyole, Dr. Nick Lyimo, for facilitating preparation of this work. The author is sincerely grateful to Mr. A.E.M. Temu for critically reading the first draft of this paper.
2. The relationship of soil organic matter (OM) to bulk density (BD) is explained by the following simple linear equation: $BD = 1.41 - 0.086*OM$; $r2 = 0.862**$ ($p<0.01$). This shows that 86.2% of the variation in bulk density can be described by variations in the soil organic matter concentration.

References

Allan, W. 1965: *The African Husbandman*. London: Oliver and Boyd.

Barley, K.P. and Greacen, E.L. 1967: Mechanical Resistance as a Factor Influencing the Growth of Roots and Underground Shoots. *Advances in Agronomy*, 19, 1-40.

Bauder, J.W., Randall, G.W. and Swam, J.B. 1981: Effect of Four Tillage Systems on Mechanical Impendence of a Clay Loam Soil. *Soil Science Society of American Journal*, 45, 802-806.

Bengough, A.G. 1991: The Penetrometer In Relation To Mechanical Resistance To Root Growth. In K.A. Smith and C.E. Mullins (eds.), *Soil Analysis: Physical Methods*, pp. 431-445. New York: Marcel Dekker, Inc.

Bengough, A.G. and Mullins, C.E. 1990: Mechanical Impendence To Root Growth, A Review Of Experimental Techniques And Root Growth Responses. *Journal of Soil Science*, 41, 341- 358.

Bowen, H.D. 1981: Alleviating Mechanical Impedance. In G.F. Arkin and H.M. Taylor (eds.), *Modifying the Root Environment to Reduce Crop Stress*, pp. 21-57. Monograph 4. St. Joseph: American Society of Agricultural Engineers.

Dregne, H.E. 1990: Erosion And Soil Productivity In Africa. *Journal of Soil and Water Conservation*, 45 (4), 431-336.

Ellis-Jones, J., Martin, L.H., Kayombo, B., Dihenga, H. and Thadei, S. 1994: *A Participatory Rural Appraisal of Mbinga District, Tanzania: With Emphasis on Existing Soil and Water Conservation Systems*. Working Document I. UK: Silsoe Research Institute.

Ellis-Jones, J., Kayombo, B., Martin, H.L. and Dihenga, H. 1996: *What Future For The Ngoro Soil And Water Conservation System: An Interim Evaluation*. Proceedings of Environment Research Project Regional Workshop (ERPRW), Nveri and Embu, Kenya, 26-29 May 1996. SRI Report No: IDG/96/15.

Gichangi, E.M., Jones, R.K., Njarui, D.M., Simpson J.R., Mututho, J.M.N. and Kitheka, S.K. 1989: Pitting Practices For Rehabilitating Eroded Grazing Land In The Semi-arid Tropics Of Eastern Kenya: A Progress Report. In: *Proceedings of 6th International Soil Conservation Organization*, pp. 313-327. Nairobi.

ICRA (International Centre For Development Oriented Research In Agriculture). 1991: *Analysis Of The Coffee Based Farming Systems In The Matengo Highlands, Mbinga District, Tanzania*. Working Document Series 15. Wageningen: ICRA.

Kayombo, B. 1992: Tillage Systems And Soil Compaction In Semi-arid Tanzania. *Proceedings of Tanzanian Society of Agricultural Engineers*, 4, 80-88.

Kimaro, D.N., Meliyo, J.L., Msanya, B.M., Magoggo, J.P. and Wickama, J.M. 1996: *Investigations of Environmental Factors for Land Management in Litembo Village, Mbinga, District, Tanzania*. Technical Report 4. Mombo Woodland Research Project, Natural Resource Study Team. Morogoro: Sokoine University of Agriculture.

Lal, R. 1986: Soil Surface Management In The Tropics For Intensive Land Use And High And Sustained Production. *Advances in Soil Science*, 5, 1-108.

Larson, W.E. 1962: Tillage Requirements For Corn. *Journal of Soil and Water Conservation*, 17 (l), 3-7.

Linden, D.R. 1982: Predicting Tillage Effect On Evaporation From Soil. In P.W. Unger and D.M. Van Doren (eds.), *Predicting Tillage Effect on Soil Physical Properties and Processes*, pp. 117-132. American Society of Agronomy Special Publication 44. Madison: American Society of Agronomy.

Lowdermilk, W.C. 1986: Man Made Deserts. In R.P.C. Morgan (ed.), *Soil Erosion and its Control*, pp. 9-19. New York: Hutchison Ross Publication.

Malley, Z.J.U. 1997: *Effect of Pit Size in* Ngoro *Cultivation System on Soil Water Conservation and Yield of Maize* (Zea mays L.). M.Sc. Dissertation. Morogoro: Sokoine University of Agriculture.

Marshall, T.J. and Holmes, J.W. 1988: *Soil Physics*. 2nd Ed. New York: Cambridge University Press.

Martin, H.L., Willcocks, T.J. and Kayombo, B. 1996: *Technical Aspects Of The Ngoro/Matengo Pit System Of Cultivation From A Soil And Water Conservation Viewpoint*. Proceedings of Environment Research Project Regional Workshop (ERPRW), Nyeri and Embu, Kenya, 26-29 May 1996. SRI Report No: IDG/96/15.

Meek, B.D., Rechel, E.R., Carter, L.M. and DeTer, W.R. 1992: Bulk Density Of A Sandy Loam: Traffic, Tillage, and Irrigation-Method Effects. *Soil Science Society of America Journal*, 56, 562-565.

Monteith, N.H. 1974: The Role of Surface Roughness in Runoff. *Journal of Soil Conservation*, New South Wales 30, 42-45.

Msanya, B.M., Kimaro, D.N. and Magoggo, J.P. 1995: *Pedolocrical Investigations and Land Resources Characterization in LuPilo village, Mbinga, District, Tanzania*. Technical Report 1. Miombo Woodland Research Project, Natural Resource Study Team. Morogoro: Sokoine University of Agriculture.

Msanya, B.M., Wickama, I.M., Kimaro, D.N., Magoggo, J.P. and Meliyo, J.L. 1996: *Investigations of the Environmental Attributes for Agricultural Development in Kitanda Village, Mbinga, District, Tanzania*. Technical Report 5. Miombo Woodland Research Project, Natural Resource Study Team. Morogoro: Sokoine University of Agriculture.

Mututho, J.M. 1989: Some Aspects of Soil Conservation on Grazing Land. In *Soil And Water Conservation In Kenya: Proceedings of the Third National Workshop*, pp. 315-322. 16- 19 Sept. 1986. Nairobi.

Onstad, C.A. 1984: *Depressional Storage on Tilled Soil Surfaces*. Transactions of the

American Society of Agricultural Engineers 27 (3), 729-732.

Romkens, M.J.M. and Wang, J.Y. 1986: *Effect of Tillage on Surface Roughness.* Transactions of the American Society of Agricultural Engineers 29(2), 429-433.

Stenhouse, A.S. 1944: Agriculture in the Matengo Highlands. *East African Agricultural Journal* 9, 22-24.

Stewart, B.A. and Robinson, C. 1996: Sustaining Soil Quality in Dryland Regions. *Desertification Control Bulletin*, 28, 24-27.

Thadei, S.Y. 1995: *Evaluation of Erosion by Modelling: A Progress Report.* Miombo Woodland Project. Morogoro: Sokoine University of Agriculture.

UAC (Uyole Agricultural Centre). 1988/89: *Annual Research Report*, Part H, No. 56. Mbeya: UAC.

Unger, P.W., Stewart, B.A., Parr, J.F. and Singh, R.P. 1991: Crop Residue Management and Tillage Methods For Conserving Soil and Water in Semi-arid Regions. *Soil Tillage Research*, 20, 219 - 240.

Zobeck, T.M. and Onstad, C.A. 1987: Tillage and Rainfall Effects on Random Roughness: a Review. *Soil and Tillage Research*, 9, 1-20.

[illegible]

CHAPTER 3

SUSTAINABLE FARMING IN A FRAGILE MOUNTAIN ECOSYSTEM: THE CASE OF CENTRAL KIGEZI, UGANDA

by Dezi S. Ngambeki, Alex R. Tindimubona and Sunday Mutabazi

INTRODUCTION

In Kigezi, southwestern Uganda, mountain ranges rise 1,500 to 2,500 metres above sea level. Such mountainous areas play an important environmental role in modulating climate, supplying water and hosting forest (UNCED, 1992). The Kigezi region is also very important to Uganda because the region contains the southwest mountain tourism zone, which includes the Mgahinga Gorillas park, a game reserve, a large proportion of the tropical high forests, significant biological diversity with species of birds, butterflies, amphibians and some endemic small mammal species in their natural habitats (Kasenene & Basuta, 1996; NEIC, 1993). But Kigezi is also a highly populated region that depends on agriculture.

During the 1940s, 1950s and 1960s, Kigezi was well known for its successful mountain agriculture based on terracing. But by the 1990s this had changed dramatically. The terraces were abandoned, soil erosion was increasing and agricultural productivity was in decline. The research project sought to examine farming systems in central Kigezi, to understand why the terraces were abandoned and to gather attitudinal information which could be used to design ameliorative public policy. It was hypothesized that there is an information gap among the generations under the age of 40 with respect to technological knowledge and mountain farming technologies available from an earlier period in Kigezi and from elsewhere.

SOIL CONSERVATION POLICY DURING THE COLONIAL ERA (1922-1960)

Department of Agriculture reports (DOA, 1940 to 1945) indicate that when the colonial administration came in the 1920s, most of Kigezi was characterized by mountain ranges with steep fluted hillsides and U-shaped valleys, some of which were wet with papyrus swamps and others dry. Traditionally, the people lived at the base of the slopes, cultivated the hillsides and used the hilltops and valleys for grazing. Before 1940, in Kigezi, the soils along the hill slopes, hollows and along foot slopes were deep, with good soil structure and a pH of 6.7. Soil was developed on phyllites, shales and sand-stones and although permeable by water, the soils were not leached (DOA, 1945). The soils on the hilltops were shallow and the sub-soils consisted of comminuted rock fragments.

In order to boost and maintain cotton and coffee production elsewhere in Uganda, the colonial government had introduced a policy of laying out land for block farming. People carried out group cultivation on plots parallel to contours and separated by planted grass strips for wash-stops as a soil conservation measure. In Kigezi, the very steep slopes were not conducive to block cultivation. Instead, a system of terracing was designed to conserve soils. Base contour lines were established at each 2-metre vertical drop and permanently marked by planting elephant grass or ficus stumps. Then the chiefs were instructed to measure parallel strips, leaving plot widths of 12 metres where the angle of slope was 15° to 20° and 16 metres on less steep slopes. In this terracing system, successive plots were separated by a three-yard planted grass wash-stop on less steep slopes, or by leaving every alternate plot under fallow and then putting trash bunds halfway up each cultivated plot on the steep slopes. The very steep slopes and hilltops were left for planting wattle trees for firewood. Then district by-laws were enacted to enforce the system of terraces with a prison sentence and/or a penalty fine.

The drought of 1943 caused famine in Kigezi and the colonial government introduced a policy of swamp reclamation. In major swamps in Kashambya and Kabale, extensive swamp drainage schemes were undertaken to create areas for planting sweet potatoes, Irish potatoes, sorghum and pastures (DOA, 1944). By 1944, the food insecurity problem was being aggravated by over-population, especially around Kabale and Rukiga, where people settled around the administrative centres. That year, the colonial government appointed a committee to study and make suggestions to address the issue. After surveys and laboratory analyses of the over-crowded soils, two main decisions were made. First, a resettlement scheme was initiated and some 80,000 people were moved from Kigezi to neighbouring areas with less population (such as Rujumbura, Rwampara, Buhweju and Igara) in western Uganda. Second, it was decided to re-enforce the terracing system with strict control on land utilization. It became compulsory to maintain hedge rows including hedged

Figure 3.1 Kigezi slopes before 1960

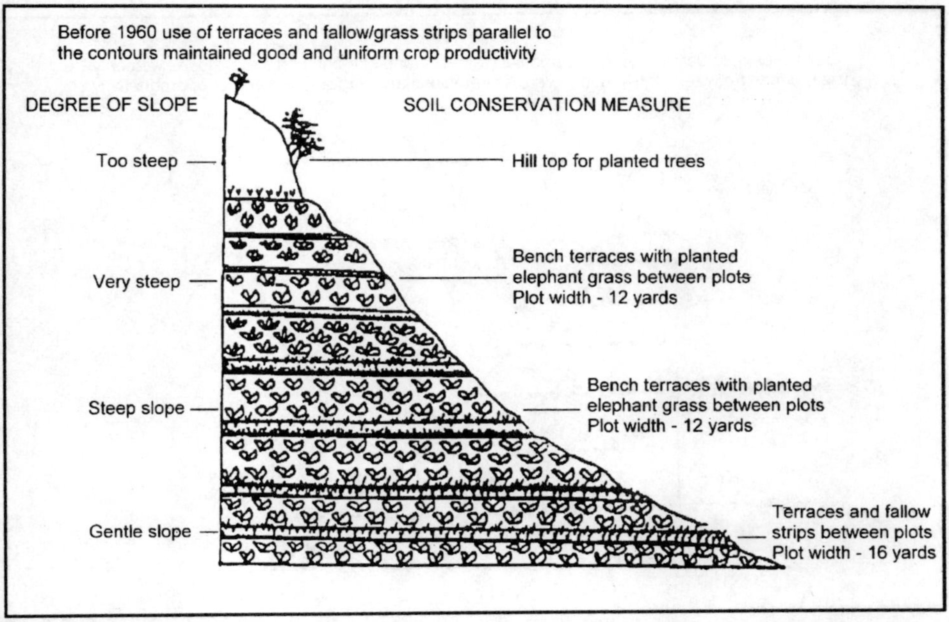

cattle tracks. Practices were enforced, such as planting black wattle trees on contour lines, back-sloping terraces on steep slopes and a rotation system in which plots were cultivated for no more than two years followed by at least four years of rest under grazed pastures. As soil accumulated, the area between the successive bunds kept leveling out and reducing the gradient. The soils were fertile and very productive. By 1960, most of the hills were covered with beautiful, almost flat green terraces. The natives of Kigezi were well known for their hard work and their skill in using intensive agricultural production methods.

But by the 1990s, this had changed dramatically. The terraces were abandoned. The soils on the mountain slopes are now in a very poor state. Most mountain ranges are bare, creating serious shortages of fuel wood and poles. Excessive soil erosion, sheet soil erosion and landslides are common. Consequently, drainage in the wetlands has been threatened and conversion for agricultural activities is difficult. Farm yields in Kigezi are now among the poorest in Africa. Yet in the Machakos District in Kenya, with a similar mountain slope farming system, farmers have almost eliminated soil erosion problems since the 1940s by building narrow-based terraces (*fanya chini*) and bench terraces (*fanya juu*) on almost all the mountain slope plots (English, Tiffen & Mortimore, 1994).

Figure 3.2 Kigezi slopes after 1960

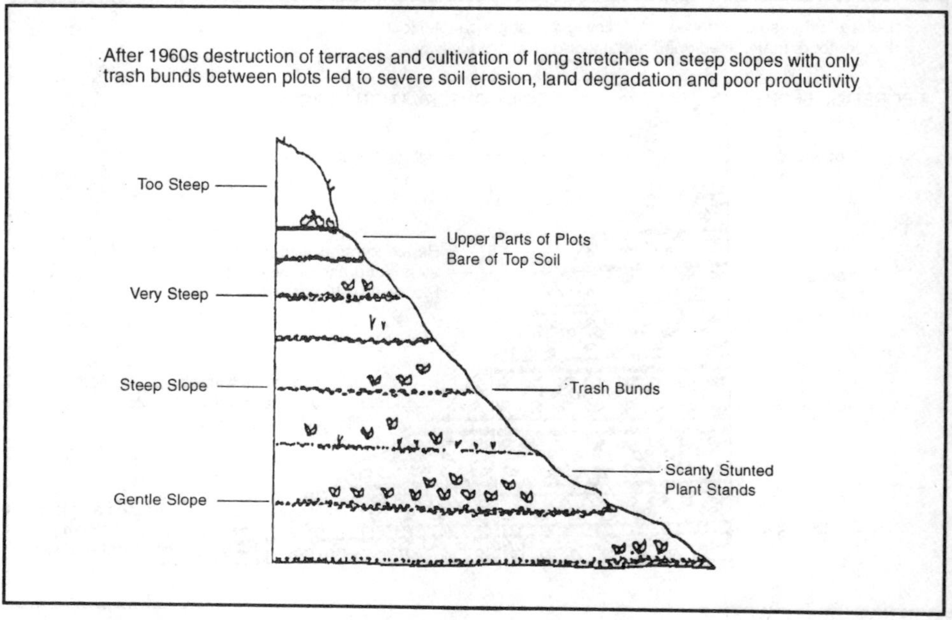

THE RESEARCH PROJECT

The research project sought:

- to examine the present constraints on the farming systems in central Kigezi by analyzing basic data on crops and livestock production systems, farming practices and the status of natural resources;
- to identify and analyze the root causes that brought about the abandonment of terracing and inappropriate usage of farm and natural resources;
- to provide information on social, economic and agricultural technological gaps, including people's perceptions and attitudes to root causes which could be used to design ameliorative public policy.

The area of study in central Kigezi consisted of four sub-counties – Kashambya and Hamurwa in Kabale district plus Nyakishenyi and Rutenga in Rukungiri district. Each sub-county is divided into smaller administrative units similar to parishes, which are in turn subdivided into villages or LC1 cells. According to the 1991 population census, Nyakishenyi has eight parishes, over 23,000 people and a population density of about 199 persons/km^2. Kashambya has six parishes, over 22,000 people and a population density of about 150 persons per

km^2. Hamurwa also has six parishes, over 22,000 people, but a population density of over 200 persons per km^2. Rutenga is the smallest, with four parishes, a population of about 12,555 and a population density of about 100 persons per km^2.

In mid-July 1997, the research team carried out a reconnaissance tour of central Kigezi. The tour covered four selected sub-counties and identified the target parishes, established personal contacts and noted outstanding characteristics in the study area. The sampling structure for a participatory rural appraisal (PRA) was completed in August 1997. The existing administrative units in each selected sub-county were used. The parishes in each sub-county were numbered. In Humurwa, Rutenga and Kashambya, two parishes were randomly selected and from those, three to four villages were randomly selected for sampling. In Nyakishenyi, five parishes were randomly selected and from each of those, five to eight villages were randomly selected. Nyakishenyi was given extra weighting because it was considered the centre of the project area, it exhibited most of the farming systems characteristics and its population density was almost twice as high as those in the other selected sub-counties. In total, 58 villages were chosen for sampling from the 11 selected parishes.

The participatory rural appraisal meetings

The PRA was conducted at the beginning of September, the findings were compiled and a report was written in October 1997. The aim of the PRA meetings was to discuss broad themes and to identify farmer priorities regarding their farms. The participants included:

- all full-time farmers (men and women);
- all livestock farmers (men and women);
- all leaders of farmer groups and social associations and civic leaders such as LCs, chiefs, government, educational and religious leaders.

One PRA meeting was conducted in each of the 11 selected parishes, and between 50 to 150 farmers attended each meeting. Each PRA meeting was chaired by a villager and facilitated by researchers using a prepared check-list to guide the discussions. In all cases, village meetings appointed their own reporters, deliberated and made resolutions about their community participatory development strategies. The researchers took note of the discussions and asked probing questions. After the plenary, farmers were divided into focus groups to discuss issues and formulate group action plans.

Some of the critical questions and issues on which there was no consensus during the meetings were followed up by the researchers with focus group or key informant interviews. This process helped the researchers improve the levels of accuracy on the obtained data.

Census survey of good farmers

After the PRA village meetings, a census questionnaire was designed. This questionnaire included questions supplied by each of the three collaborating projects. In October 1997, a listing of good farmers from each of the selected parishes was compiled. The criteria for selecting the good farmers were:

- the size of their farm
- their full-time and active engagement in farming
- the number of years they had been farming
- evidence of healthy, well-managed crops and/or livestock

Then 5 to 15 farmers were randomly selected from the parishes under study, making a total sample of 120 farmers for the baseline survey. At the end of November 1997, the baseline questionnaire was designed and pre-tested, and the farmers were selected. Before the baseline data collection, six research assistants were recruited and trained for eight days in interview methods and field data collection. Two were assigned to each sub-county and regularly supervised by the researchers.

The detailed baseline data was collected from December 1997 to February 1998. The aims of the census survey were to obtain broad-based farmers' perceptions of key issues and to quantify those issues. The census survey targeted the same 58 villages chosen from the 11 selected parishes. The researchers had decided to collect the data during the peak period of agricultural activities, September to December, which is normally a short rainy season. However, in 1997, that period was excessively wet due to the El Niño phenomenon. The heavy rains made it impossible to travel to a few of the selected villages, so they could not be included in the baseline survey and data could not be collected.

According to the returns from the census survey, 434 good farmers were interviewed from five parishes selected from Nyakishenyi sub-county and the two selected parishes from each of the other sub-counties (Hamurwa, Rutenga and Kashambya). After the field data collection, the questionnaire responses were verified for omissions and accuracy. Then the questionnaires were coded and local measurements were converted to standard measurements. The analysis was carried out using SPSS to get descriptive statistics and multiple regressions results. Data analysis and draft report writing were carried out in February/March 1998.

Results

Demographics

The average age of the farmers was found to be 42, within a range from 15 to 80 years. The majority (60.5%) of the farmers were between 15 and 45 years old, suggesting that those farmers had not been directly influenced by the colonial regime because they were below 10 years of age in 1962, when Uganda gained independence. The average household was made up of 6 persons, with a range from 2 to 21 persons. On average, the farmers had 6.2 years of education; about 15% of the farmers had no education (Table 3.1).

Table 3.1 Demographic characteristics of good farmers from Central Kigezi, 1997

a) Summary Statistics	**Mean**	**Minimum**	**Maximum**		
Age	42.0	15	80		
Family size	6.0	2	21		
Education in years	6.2	0	15		
Other training in years	0.5	0	3		
b) Percentages per age group					
Age group, years	15-30	31-45	46-60	61-80	
Percentage	25%	35%	5%	24.9%	8%
c) Percentage per size of family group					
Family size group	2-5	6-10	11-15	16-21	
Percentage	39%	43%	7.1%	1.6%	
d) Percentage by level of education					
Level of education	None	Lower Primary	Upper Primary	Secondary	Tertiary
(Years of education)	(0)	(1-3)	(4-7)	(8-11)	(12-15)
Percentage	15%	9.5%	47.2%	20%	8.5%
e) Percentage by sex					
Sex	Male	Female			
Percentage	86%	14%			

Farm resources

The farmers in Central Kigezi live in poverty. They are engaged in subsistence farming, some with no farm resources and others with very few. On average, each household had 2.97 acres of land, within a range of no land to about 40 acres. Only a few farmers owned more than 11 acres of land. Land fragmentation was common. The average farmer owned land in many locations and planted about 2.5 acres under annual crops, but these were divided up into 4 to 5 scattered plots of about 0.5 acres each. The average farmer owned one to two cows, two to three goats and a sheep, as well as small farm tools, typically one cutlass and two hand hoes. A quarter of the farmers earned less than US$20 per annum, and 73.2% earn an annual income of less than US$200, which is inadequate for the size of their families. The farmers in the study area do not have a viable and reliable source of income. They lack a good cash crop – the average farmer receives small amounts of revenue from as many as 18 different sources. The annual income was, on average, US$190 per household. But household income ranged from zero to as high as US$4,200 per annum (Table 3.2).

Table 3.2 Farm resources owned by good farmers from Central Kigezi

a) Summary Statistics	**Mean**	**Min.**	**Max.**			
Farm size (acres)	2.97	0	40			
Cows number	1.3	0	25			
Goats number	2.8	0	28			
Sheep number	1.6	0	20			
Annual income, US$	190	0	4200			
b) Farm Size (acres)	0-0.5	0.6-2	2.1-5	5.1-10	11-20	20 +
Percentage	14.7%	15.4%	44%	15.7%	5.5%	1.8%
c) Annual income (US$)	0-20	21-200	201-500	501-100	1001-4200	
Percentage	24.4%	48.8%	17.7%	6.7%	2.9%	

Destruction and abandonment of terraces

Visits to farmers' fields showed that many farmers had abandoned terracing as it had been practised in the colonial era. Instead, they cultivate the entire hillside, putting small trash bunds at irregular intervals. There were indications that the soils are progressively being dragged down the slopes during cultivation of millet and sweet potato crops.

According to the PRA results, the farmers destroyed terraces in the late 1960s, in order to:

- access the good topsoils that had been trapped and accumulated on the terraces,
- get rid of very high benched terraces which had become too high for people to climb over,
- get rid of rodents which were multiplying in the terrace strips of bushes and grasses.

Farmers chose not to construct new terraces after destroying the old ones because it required too much labour and because of the high population pressure and land scarcity.

It should be noted that there had been problems with the terrace design in Kigezi during the colonial era. The alternate strips that were left fallow between successive cultivated plots were not wide enough to allow animals to graze without disturbing the crops in the surrounding plots. In addition, the fallow plots were not cultivated to fodder crops. Suitable trees or grasses were not planted on bunds or terraces. The colonial administration planted elephant grass on base contours, but this had disadvantages including soil nutrient depletion. The wattle trees also were not compatible with the Kigezi cropping systems, a weakness which colonial reports acknowledged. These were disincentives to the maintenance of the terracing system.

Current attitudes

Communities now set priorities according to majority views. This is a significant change from the colonial era, when constructing terraces and planting hedges were compulsory. Although the district soil conservation by-laws were not repealed, there is now no clear enforcement agent. Most of the farmers are not aware of the far-reaching implications of the land degradation caused by abandoning the terraces and other suitable soil conservation measures.

The survey revealed that the majority of farmers were not aware that fertilizer use increases productivity. Many were not aware of the significance of leaving alternating fallow plots on the hill slopes. Almost one-third were not aware that crop rotation is important and over one-third were not aware that continuous monocropping on the same piece of land may adversely affect farm productivity. Many of the farmers were not aware of the effects of terracing on hill slopes, drainage of wetlands, deforestation and control of soil erosion (Table 3.3).

Table 3.3 Percentage of farmers not aware of some factors/practices affecting productivity and environmental degradation

Factor	Percentage of farmers (not aware)
Fertilizer use	72.6
Alternating fallow plots on hill slopes	42.3
Crop rotation	29.3
Terracing on hill slopes	22.1
Drainage of wet lands	63.4
Deforestation	27.2
Control of soil erosion	10.0
Continuous monocropping	37.2

Table 3.4 shows how farmers explained soil erosion. More than half believed that rampant bush burning was the major cause of soil erosion, about one-third recognized that continuous cultivation was a cause and about one quarter recognized that lack of tree planting was a problem. But only 12.8% attributed soil erosion to the destruction of terraces on hill slopes. Even fewer recognized that cultivating hill slopes without leaving alternate fallow plots and allowing over-grazing cause soil erosion.

Table 3.4 Farmers' attributions of root causes of soil erosion.

Factor	Percentage of farmers
Bush burning	56.9
Destruction of terracing on hill slopes	12.8
Stretches of cultivation on hill slopes without alternate fallow plots	8.1
Continuous cultivation of the same land	34.4
Over grazing on same land	9.3
Lack of planted trees	26.7

Many of the old traditional farming practices in Kigezi have not changed. Methods such as cultivating millet and sweet potatoes on the slopes and designating certain pieces of land because of its location, soil type or level of fertility for the cultivation of specific crops like sorghum, sweet potatoes or Irish potatoes have perpetuated monocropping. Land reclamation, introduced to create more land to grow crops, and bush burning to regenerate young and tender grasses for grazing are all environmentally unfriendly practices.

Multiple Regression Analysis:

A multiple regression analysis examined the factors (positively or negatively) affecting farm productivity under the present farming practices. The level of farm income was used as the dependent variable and taken as a proxy for farm productivity. The set of independent variables included demographic, farm resources, farming practice and weather variables.

$Y = f(x_1, x_2, x_3......xn) + e$

Where:

(i) Dependent variable

y = level of annual farm income in Uganda shillings (from both crop, livestock and non- farm incomes)

(ii) Demographic variables

x_1 = age of the farmer in years

x_2 = educational level, years of schooling

(iii) Farm resources variables

x_3 = farm size, land holding in acres

x_4 = number of cows owned

(iv) Farming practices

x_5 = dummy for terracing yes = 1 and no = 0

x_6 = a dummy for soil erosion control yes = 1 and no = 0

x_7 = a dummy for climatic change; 1 = unexpected climatic change; 0 = climatic change.

Several regression equations were fitted and the equation of best fit is presented in Table 3.5. The regression coefficient of educational level was positive and highly significant at 1 percent. Then the regression coefficient of cows, and terracing, were also positive and significant at 5 per cent. The coefficient of climatic change was negative and significant at 5 per cent, whereas the coefficient of soil erosion control was negative but not significant. The regression coefficients of farmer's age and farm size were positive but not significant.

Table 3.5 Multiple regression analysis of factors affecting farm productivity

Dependent variable-Annual farm cash income.		
Independent viables	B coefficient	T Statistics
Education level	47.70	6.293 **
Cows	8.187	1.634 *
Terracing	106.805	1.442 *
Age of farmer	2.517	1.132
Farm size (acres)	1.981	0.964
Climatic change	-186.088	-2.470 *
Soil erosion	-111.853	-1.097
Constant	-26.056	-1.64

** Significant at 1 percent
* Significant at 5 percent

The multiple regression analysis suggests that education level (which represents skills, creativity, initiative, capacity for management) and terracing (which may also represent proper land management/soil conservation) are the major factors that could contribute positively to the improvement of farm productivity in Kigezi. The positive sign and level of significance of the included variable, COW (which represent livestock) in the regression analysis suggests the importance of mix (crop-livestock) farming. But the non-significant coefficient of the variable LAND suggests that some farmers with large farms may not be using their land effectively, with the result that the bigger holdings are not necessarily highly productive. The negative sign and significant coefficient of the terracing dummy confirms that destruction and abandonment of terraces adversely affect farm productivity.

Dissemination Workshop

A dissemination workshop was organized in April 1998 to share the research results with participating farmers, extension agents, LCs, opinion leaders and other decentralised district government officials. The farmers appreciated the research output and they became aware of the possible factors contributing to farm productivity and, most of all, the dangers and causes of land and environmental degradation in their area. The farmers' major responses to these findings were:

- recognition of the need for improved education, information, communication and extension services, and
- the importance of soil and environmental conservation, especially reviving the use of terraces.

Conclusion

In order to maintain productivity, a mountain farming system must retain top soils on the slopes. The research results strongly suggest that the major practices causing land and environmental degradation in Kigezi are:

- destruction of terraces along the hill slopes;
- rampant bush burning;
- abandonment of alternate fallow strips and plots when cultivating on hill slopes;
- continuous cultivation on the same piece of land, without use of fertilizer or fallowing;
- wetland drainage for agricultural activities; and
- deforestation without active tree planting.

The research demonstrated that the farmers' knowledge base is weak. There is great need to sensitize farmers on the urgency of proper soil conservation. Making the younger generation aware of the consequences of unsuitable farming practices and providing information on the available options could bring about a dramatic change towards developing a successful mountain farming system in Kigezi. There is a need to revive district and local government by-laws to stop bush burning and enforce the use of terraces on mountain slopes.

But soil conservation alone is not sufficient. Productive farming requires quality farm labour, successful crop combinations, a balance between crops and livestock and appropriate technological inputs (Wilson & Kang, 1981; Ngambeki & Ikpi, 1982). To date, little attention has been given to developing a farming system that is suitable for the Kigezi region. There have been some unsuccessful searches for suitable cash crops, ranging from black wattle trees, flax, tobacco, castor plant to pyrethrum (Bazirake, 1991; DOA, 1948, 1957). But these attempts have overlooked the importance of dual purpose (cash and food) crops, such as beans and bananas, which farmers in other parts of South Uganda have successfully adopted. They have also overlooked the need to integrate into the farming system the farmers' traditional farming practices and the already adapted crops such as field peas, Irish potatoes and sorghum, which are typical crops in the region. Thus, more research and farmer extension efforts will be required to establish a renewed mountain farming system for the Kigezi region.

References

Bazirake, C. 1991: *The search for a cash crop for Kabale District.* A paper presented at a seminar for Kabale District Development. Unpublished.

English, J.M., Tiffen, M. and Mortimore, M. 1994: *Land Resource Management in Marchakos District, Kenya 1930-1990.* Washington D.C.: World Bank.

Kasenene, J.M. and Isabirye Basute, G. 1996: *Summary of Existing Information on Forestry, Protected Areas and Terrestrial Bio-diversity Conservation in Uganda.* Unpublished.

DOA (Ministry/Department of Agriculture). 1940, 1944, 1945, 1948, 1957, 1970, 1972: *Annual Reports.* Entebbe: Uganda Government Printer.

NEIC (National Environment Information Centre). 1993: *Policy Profiles, Biodiversity in Uganda*. Kampala: Ministry of Natural Resources.

Ngambeki, D.S. and Ikpi, A.E. 1982: Rural Farmer Productivity in the Tropics: The Nigerian Experience. *Canadian Journal of Agricultural Economics,* Vol. 30 (3), 297-318.

UNCED. 1992: *Rio-Declaration on Environment and Development.* Rio de Janeiro: The United Nations Conference on Environment and Development.

Wilson, G.F. and Kang, B.T. 1981: Developing stable and productive biological cropping systems for the humid tropics. In B. Stonehouse (ed.), *Biological Husbandry: A Scientific Approach to Organic Farming.* London: Butterworth.

CHAPTER 4

CROPPING GUIDELINE FOR THE MACHOBANE FARMING SYSTEM, LESOTHO

by Letlamoreng Mosenene[1]

INTRODUCTION

The Machobane Farming System has been designed to produce several crops on a piece of land in a relay and intercropped fashion over a year. It uses farmyard manure or compost and wood/dung ash to manage soil fertility. Compared with the common practice in Lesotho of repeated maize growing every year, the Machobane system offers poor farmers many advantages:

- **Food security through crop diversification.** When drought, flooding, hailstorm, frost, pests strike, they cannot wipe out a whole year's produce.
- **Improved diet and health.** When other crops are available in abundance throughout the year, the families of poor farmers can eat more and better food all the time.
- **Improved soil and water conservation.** Ploughing is not done every year, manure and plant compost are returned to the soil, legumes are included in rotations, fields are covered throughout the year; humus builds up in the soil and there is better water retention.
- **Increased food production.** When limited nutrients are concentrated in a single field which is looked after very well, and many successive crops are grown in a year, yields from that single field can exceed that of four to five fields that are less intensively cropped.
- **Limited required inputs.** Manure, ash and sometimes sand are concentrated in one field and applied only in furrows where crops grow, rather than in the aisles, where they are more likely to fertilize weeds. There is no need for inorganic fertilizer use. When farmers work in the field every day as they should in this system, pests are not allowed to reach exploitative levels, and therefore the use of chemicals is very limited.

BACKGROUND

The system was developed by James Jacob Machobane. When he left regular employment and took his family to live in Nqechane, Leribe, Lesotho, he found he had to make a living from the soil. He was disturbed by the fact that when crops failed, Basotho farmers tended to adopt a fatalistic attitude. They would pray to God, and would do nothing to help themselves (Lewis, 1959). He maintained this attitude has "prolonged the dependency which has led to abject poverty and hunger of the day, the starvation of no less than of true slavery...there is need to hasten the ending of human apathy and indignity...saving victims from perpetual foreign charitable aids" (Machobane, 1981).

Machobane ascertained that the majority of Basotho who owned land were subsistence farmers, who were supposed to be taught scientific methods of agricultural production by agricultural demonstrators. The main method being promoted was, and still is, that of monocropping maize, the staple food. Secondary crops included sorghum, beans, wheat, peas and pumpkins. Potatoes were relegated to horticultural status and commonly grown in gardens rather than fields. In this system of monocropping, the land is terribly under-utilized. There was no effort to make full use of the growing season in any given calendar year. There was always a high risk of failure if the season brought drought, too much rain, early or late frosts, pests or disease. Crop rotation was not practised as often as it should have been.

Consequently, the yields per acre were very low, and these have fallen further over the years. Since the mid 1970s, maize yields have declined on average from 1,359 kg/ha to 590 kg/ha in the mid-1980s and to 326 kg/ha in 1992. Crop failure for the lowlands has increased from 8% in the mid 1970s to 38% in 1991/92. The erratic climate that is becoming a trend in Lesotho will continue to put more stress on agricultural productivity.

The second biggest constraint to agriculture has been low soil fertility. Even in the 1940s, the application of kraal (farmyard) manure to maintain or improve humus was minimal. Instead, inorganic fertilizers were promoted as soil amelioration alternatives.

BASIC CROPPING PATTERNS OF THE MACHOBANE FARMING SYSTEM

Machobane developed his own set of alternative strategies and has verified them in years of experimentation (Pfotenhauer, 1987). The system is based on three principles:

- rotation
- relay planting
- intensive intercropping.

The idea is to make full use of the growing seasons, spread the risk somewhat and harvest more than one crop. Mixed intercropping, that is, taking crops belonging to one sowing period and mixing them in one row, increases variety and reduces risk. Sequential planting, that is, planting succeeding crops immediately after the preceding crop has been harvested, increases land use efficiency. The goal is to increase food production in both the vertical (yield per crop) and horizontal (yield per year) dimensions. One sample planting pattern is illustrated in Figure 4.1.

Figure 4.1 Sample planting of crops with the Machobane Farming System

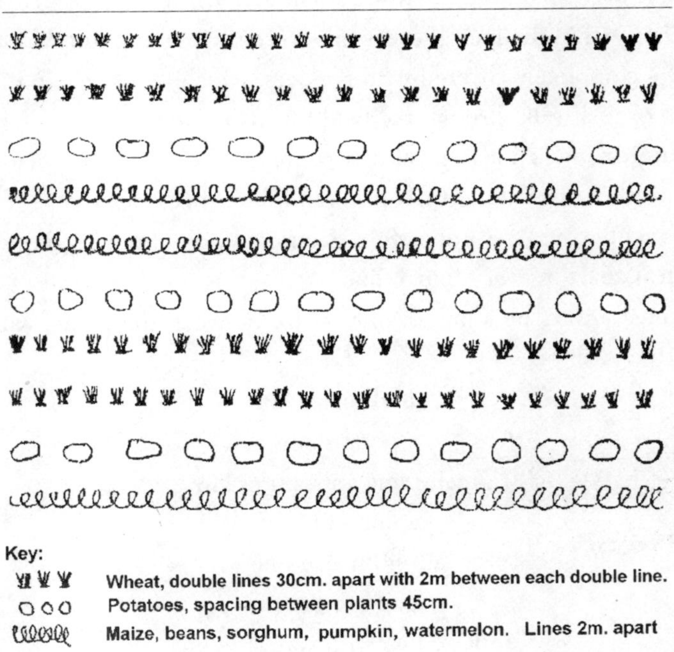

Key:

Wheat, double lines 30cm. apart with 2m between each double line.

Potatoes, spacing between plants 45cm.

Maize, beans, sorghum, pumpkin, watermelon. Lines 2m. apart

- **Potatoes** are the mainstay of the system. Potatoes are planted in lines spaced two meters apart. These lines should always run along the contour, to ensure that when ridging takes place, water erosion is minimized. Within each line, the potato pits or small holes should be 40 to 50cm. apart. The best variety, in terms of yield and resistance to both early and late blight, is BP1. The variety also is popular with potato chip makers. Normally, no other crops are sown in the potato lines.
- **Maize** is sown between the potato lines. The maize rows are one metre from the potato line and two meters apart. The spacing between plants within the maize line is 40 cm, as in regular maize planting.
- **Beans** are sown in the same line as the maize. Bean plants are spaced 20 cm apart.
- **Sorghum** is also sown in the same line as the maize and beans, at the regular spacing of 15 cm between plants.
- **Pumpkin** is sown in the same line as the maize, beans, and sorghum, but only on every fourth maize line. Because pumpkins send out runners, the seeds (two/three seeds per hole to ensure germination) must be placed at 4m from each other within the line. If every line were sown with pumpkin, the ground would be completely covered with runners, which would make it difficult to dig, hoe or just walk through the field for inspection.
- **Watermelon**, which also produces runners, is sown in the maize line, but never in the same line that has pumpkin, because access for cultural management would be impossible. Consequently, three seeds per hole are placed 4m apart in every fourth line.

  There must always be a line of only maize between the plants with runners. The pattern will be:

  a solid maize line
  a potato line
  a maize line with pumpkin
  a potato line
  a solid maize line
  a potato line,
  a maize line with watermelon
  and so on.
- **Wheat** is sown in double lines, 30 cm apart. The double lines are spaced two

meters apart to allow for the sowing of potatoes when the wheat is harvested. Figure 4.2 illustrates in a simplified manner the sequencing of sowing and harvesting for each of the seven crops.

Figure 4.2 The Machobane Cropping Pattern

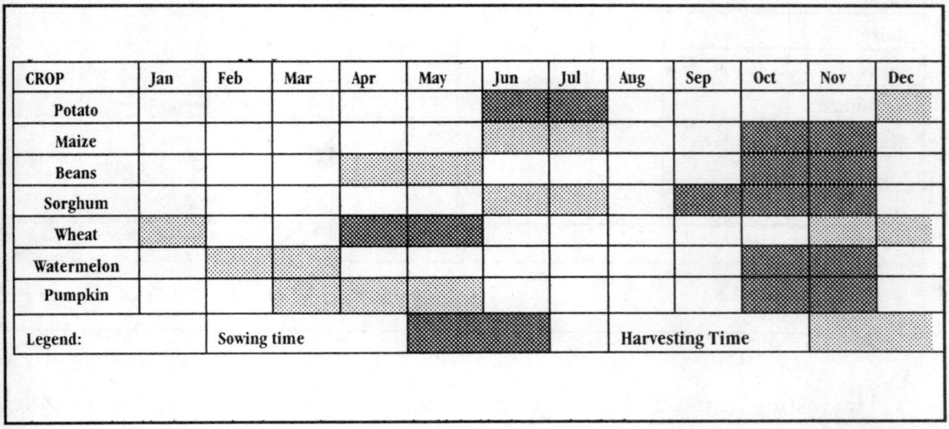

The planting dates for all crops are actually more extended than suggested above. Figure 4.3 indicates the range of planting times. The darker shades indicate the optimum planting dates for the lowlands and foothills, according to the Production Guidelines Volume II, produced by the Agricultural Research Division, and published by the Agricultural Information Services in 1991. The crosshatched areas represent marginal periods for each crop.

Potatoes require an optimum temperature of 16 C, with a range between 7 and 24 C for best production. Short days and long nights favour tuber formation and flowering (Makhata, Marlowe, Campbell & Jobo, 1991). Potatoes can be sown from mid-July to November, with optimum yields expected when the planting date is between October and November (Makhata *et. al.*, 1991). However, if the later planting dates are used, the harvest will take place at the peak time of supply to the market and the sale prices will be low. The highest prices can be obtained from September through January. July to September are sub-optimal periods for potato in areas where frosts may set back growth, but this will not kill the crop. Maize, beans, sorghum, watermelon and pumpkin all share similar sowing periods and can thus be planted in intercropped lines. Wheat can be sown from March through June.

Figure 4.3 Planting dates

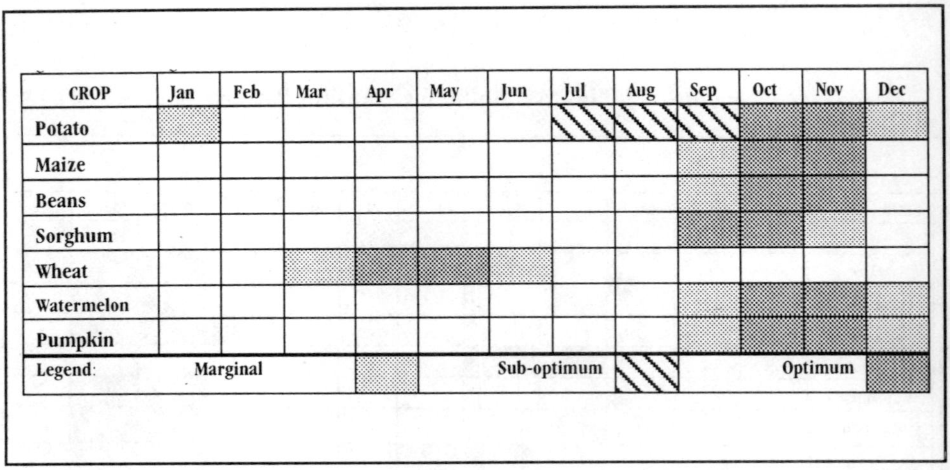

Harvesting can be done over even longer periods. Figure 4.4 includes harvesting times for crops planted during marginal dates. Potatoes can be harvested over nearly the entire year if they are kept in the ground for storage.

Figure 4.4 Harvesting dates*

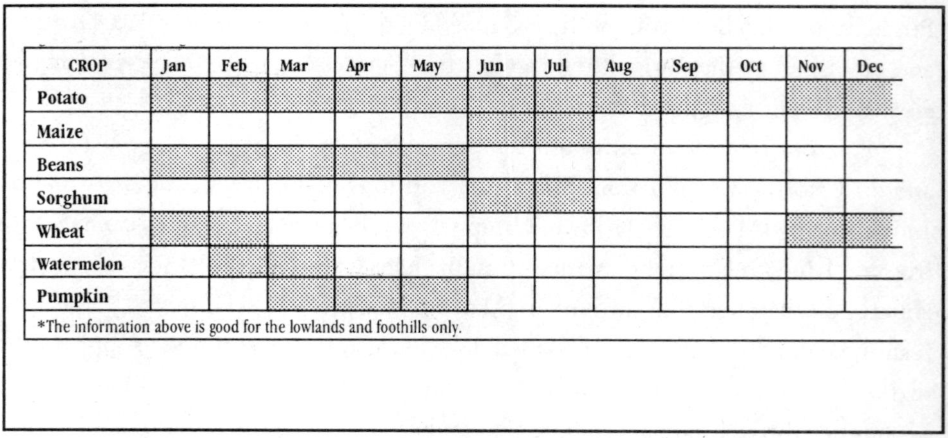

Farmers are encouraged to use open pollinated varieties because these normally have high tolerance for various crop problems brought about by weather, pests and diseases. This tolerance is a consequence of adaptation to local conditions. But, it is also acceptable to use improved varieties if the farmer has good management skills.

Cultivation Practices

Seed bed preparation

The initial seed bed preparation should be thorough, so the soil is properly porous. Deep ploughing is an advantage, as it will allow for better root penetration and increased moisture retention. Where it cannot be avoided, tractor ploughing can be done, but farmers can use oxen just as well. Further improvements can be achieved by ripping the field. Discing and/or harrowing may be necessary. Preparation of the seed bed in autumn or early winter (during the months of April and May/June) is advantageous because:

- Turning under the summer weeds increases the organic matter content of the soil.
- It allows for more moisture to accumulate for spring growth.
- Pest eggs will be killed by exposure, reducing the potential for pest outbreaks. If the seed bed is properly prepared, the soil will stay in good condition until October for summer sowing. For soil fertility maintenance, compost or kraal manure and ash are deemed sufficient. Kraal manure or compost and wood/dung ash must be hauled to the field to await the sowing of potatoes. The application rates of manure and ash are determined according to soil type as follows:
- **Coarse sand soils:** 3 scotch carts at a ratio of 2:1, manure to ash per acre.
- **Sandy soils and the duplex soils:** 4 scotch carts at a ratio of 1:1, manure to ash per acre
- **Heavy loam soils:** 3 scotch carts at a ratio of 1:2, manure to ash per acre.

The crop requirement of nutrients is satisfied by these amounts. There is no precision in terms of weight, but as a rule of thumb, the less the application, the poorer the yield will be. Toxicity due to excess application has not yet been recorded. The ratios above are good for all Machobane system crops. However, for an excellent quality watermelon, 20 litres of very fine, silty river sand should be added to the normal mixture of 1 to 2 ratio (wheel barrow) of manure to ash.

For the potatoes, two to three hand scoops of the manure/ash mixture are placed in the sowing hole and thoroughly mixed with the soil. A single seed potato is dropped into the hole; it should be well covered and sit toward the bottom of the mixture. Recently, the farmers have adapted these scoops to equal two spadefuls of the mixture in order to achieve uniform application, especially if the activity is done collectively. Similarly, the pumpkin and watermelon are sown using the prescription outlined above; the seeds are then sown in small pits toward the bottom of the mixture. For best germination results, maize, beans, sorghum and wheat prefer soil to be the seed bed, so these seeds are set in direct contact with the soil. The manure/ash mixture is applied on top, then covered with soil to prevent it being carried away by wind or water.

Weeding

It is recommended that no weeds be allowed to grow alongside crops. In addition to competing with crops for light, water and nourishment, weeds attract the ladybird which easily transmits the potato viral diseases. A weeding regime has been developed especially for the potatoes.

1) The first weeding is done soon after the emergence of the potatoes to soften the soil around the young plants, increase aeration, and to attract as much "vaporised moisture" as possible into the ground.
2) The second weeding is done when the potatoes are about one month old for the same reasons as outlined in (1).
3) The first earthing follows a couple of weeks later, when the potatoes are at the flowering stage. A very small amount of soil is gathered around the plant in acknowledgment that the tubers should be forming soon.
4) The second earthing is done when the plants are at second budding. A bit more soil is ridged. Care is taken not to overburden the young plants because that discourages tubering and encourages vegetative growth above ground.
5) The third earthing, at third budding, is quite heavy because the plants are quite big and some of the earlier formed tubers do require a lot of soil. The soil is heaped around the plant, covering almost half of it.
6) The fourth earthing is the heaviest. Only one third of the plant is left showing. No earthing will take place again because at this time the potatoes are nearing the end of their life. They will dry somewhat, which indicates they are ready for harvesting.

After every rainfall, hoeing is encouraged, to break the crust and repair the potato ridges. The best tool for ridging is a hoe and/or potato ridger and not a spade or plough. The spade or plough invariably adds more soil than necessary.

The other crops including the double rows of wheat are weeded ordinarily.

Disease and pest control

Machobane emphasizes that life forms should not be destroyed because of their role in the ecological pyramid. If chemicals are used to control disease, one might be creating an environment that would eventually render the chemicals ineffective – or create space to be occupied by a worse pest. Secondly, the chemicals may be harmful to humans as well as to the environment.

In the management of plant diseases, the use of healthy seed should be the first priority. Healthy plants in fertile soil are not usually susceptible to diseases. The constant presence of the farmer for various cultural activities should also assist in early observation of insect and disease infestations.

The Machobane system offers an opportunity for integrated pest management. Crop rotation is very important. With monocropping, areas are planted exclusively to one crop year after year, inviting severe pest damage. Pests numbers can explode because there is less interference or competition. Alternating crop species over time has long been known to lessen the incidence of pests. When a pest is unable to find a host, its population is reduced. Rotation is also effective in lowering the population of soil-borne diseases. Some of these may have resistant stages, such as spores, which may carry them through periods of no host, but in general, lack of a host reduces the amount of inoculum available for a disease outbreak the next time the same crop is planted (Litsinger & Moody, 1976).

However, crop rotation should be carefully planned and it remains a challenge for the Machobane system to develop effective pest and disease control regimes. Data on pests are usually inadequate because pests and diseases are highly dynamic. They adjust, sometimes very quickly, to changing hosts and conditions. Less pest and disease data exist for multiple cropping or intercropped environments. Experience with the Machobane system, from 1991 to the present, suggests that only the early and late blight (*Alternaria solani* and *Phytopthora infestans*, respectively) are of serious concern. These diseases may wipe out the entire potato crop within days of attack. Both can be overcome by either very early or late planting – in June/July or November/December. Otherwise, farmers should adopt a strict management regime as recommended by Makhata *et. al.* (1991): weekly spraying with maneb or antracol. Black scurf (*Rhizoctonia solani*) is a fungal disease that attacks the stem and tubers, girdling the young stem at emergence; it can be carried by mature tubers into the next growing season. It is controlled by dipping the tubers in maneb before sowing. Common scab (*Streptomyces scabies*) is soil borne. It is characterized by pitted spots on the tuber. Low pH (5.0-5.5) makes it less active. Reference for further information should be made to the Production Guidelines, Vol II, prepared by the Agricultural Research Division in 1991.

Harvest, storage, marketing

Potatoes are the most important commercial element of the system and they require particular care. The highest tuber enlargement takes place after the last earthing, perhaps because the process of senescence is in progress. When the plants or vines become yellow and die down, the tubers are becoming ready for harvesting. They should be harvested when the skin does not slip off when rubbed. Potatoes (BPI) normally require at least 150 days from emergence to maturity. However farmers can start selective harvesting 90 days after germination, because tubers will be enlarging. If sowing was done in July, this would be late October, when the food supply is diminishing and people are not getting adequate nutrition. The name Mant'sa Tlala (which means "a movement to dispel hunger") was given to the Machobane farming system because it allows for early selective harvesting. Recorded performance of a single vine in 1995 fluctuated between 1 kg and 5 kg, with an average of 1.6 kg.

During lifting, care must be taken not to bruise or cut the tubers or they will rot very easily. During this period, fresh sprouts should be nipped off, as they lower the quality of the produce. For lifting, ordinary hoes and digging forks work the soil best and cause less damage than spades and ploughs. Damaged potatoes should be set aside for immediate eating.

To increase the shelf life (or usable market life) of potatoes, they can be cured by covering them with a moist cloth or sisal bag in a warm but dark room for about two weeks. A tough, fatty layer forms beneath the skin. This way potatoes do not shrivel or bruise or even rot easily: the shelf life of fresh harvests can be increased from 3 or 4 weeks to 12 weeks. Constant inspection is required to remove the rots and green tubers. Always allow for lots of fresh circulating air during storage and ship out the oldest harvest first.

Packaging is important in the marketing of potatoes. Paper pockets of 10 or 15 kg capacity are the commonest packaging materials. Plastic mesh can be used, but to prevent greening, care must be taken that the potatoes are not in direct sunlight. During shipment, the bags must be stacked firmly, but not too tightly, as that will cause bruising. Potato prices are getting better by the year. In the highlands, especially in Thaba Tseka, the prices are twice the Maseru prices. The market is particularly good due to the Lesotho Highlands Development Project, which absorbs most of the produce. The market also responds very well to periods of drought with the highest prices realized after a drought period.

Net Return on Potatoes

Although watermelon, pumpkin and beans can also be sold, potatoes will be the most important market crop. Indicative costs and returns for potatoes are presented below. It must be noted that these are only estimates based on the farmer's own information, so the figures are variable. The fact that farmers pay one another in kind for certain operations, such as hoeing and harvesting, compounds the problem of giving a realistic cost estimate. Again, in real terms, most farmers are not paying for farmyard manure or ash because these are waste materials that are not used for anything else. To complete the costs of operations for the cropping pattern, costs for seed and planting of the crops must be included. It must be noted that the cultural activities like hoeing remain the same, that is, when hoeing is done the entire field is covered. Table 4.1 gives a breakdown of costs.

Financial Analysis

Crop: potatoes Variety: BP1

Management: Low input but intensive management

Area: 1 acre Market: Maseru

Average yield: 2,500 kg or 250 bags per acre

Price / bag: M12.76

Gross Income: M3, 190

Variable cost = M 848.50

Fixed cost = M 304.05 (Fixed assets are depreciated by 15% on 1995 prices for a plough (M200), scotch cart (M1, 500) 3 yokes (M300), a digging fork (M24), and a hoe (M13)

Net return = M 2, 037.95 per acre

Net return[2] will vary significantly with prices and yield levels. In the last five years, prices have varied from a low monthly average of M6.87 to a high monthly average of M18.33. In Table 4.2 the lowest and highest monthly averages are used, as well as the monthly averages for the months of January to April 1996 (M15.06, M12.91, M12.08 and M12.40). Yields variations from 50 bags per acre to 500 bags per acre are included.

Table 4.1 Operational costs for potatoes

Operation	Number of Items	Unit Cost	Cost / acre
Tillage (oxen)	1 ploughing	M 40.00	M 40.00
Manure application		Per scotch cart	
Manure	1	M 30.00	M 60.00
Ash	1	M 10.00	M 20.00
Hauling	1	M 14.00	M 56.00
Application	1 (by a person)	M 10.00 / 8 hrs	M 10.00
Planting			
Seed	1	M 50.00 / 30 kg	M 200.00
Alignment & pitting		M 10.00 / 8 hrs	M 5.00
sowing		M 10.00 / 8 hrs	M 5.00
Weeding	2	M 5.00 / 4 hrs	M 10.00
Earthing	4	M 5.00 / 6 hrs	M 50.00
Disease control			
Local concoction	4	M 5.00	M 5.00
Harvesting			
Lifting	1	M 10.00 / 8 hrs	M 105.00
Sorting	1	M 10.00 / 8 hrs	M 20.00
Bags	250	M 57.00 / 100	M 142.50
Hauling	1	M 30.00 / load	M 120.00
		TOTAL	**M 848.50**

Table 4.2 Net returns on potatoes at various yields per acre and various prices

Yield *	Prices per bag (10 kg) in Maluti					
	6.87	12.08	12.4	12.76	12.91	15.06	18.33
50	-808.5	-532	-532	-514	-506.5	-399	-235.5
60	-739.9	-427.3	-408	-386	-377.4	-248.4	-52.2
70	-671.1	-306.4	-284	-258.8	-247.7	-97.8	**131**
80	-602.5	-185.7	-160	-131	-119.2	**52.8**	**314.4**
90	-533.8	-64.85	-36	-3.6	**9.9**	**203.4**	**497.7**
100	-465.1	**55.95**	**88**	**124**	**139**	**354**	**681**
110	-396.4	**176.75**	**212**	**251.6**	**268**	**504.6**	**864.3**
150	-121.5	**659.95**	**708**	**762**	**784.5**	**1107**	**1597.5**
200	**221.95**	**1264**	**1328**	**1400**	**1430**	**1860**	**2514**
250	**565.45**	**1868**	**1948**	**2038**	**2075.5**	**2613**	**3430.5**
300	**908.95**	**2471.9**	**2568**	**2676**	**2721**	**3366**	**4347**
375	**1424.2**	**3377.9**	**3498**	**3633**	**3689.3**	**4495.5**	**5721.8**
450	**1939.5**	**4284**	**4428**	**4590**	**4657.5**	**5625**	**7096.5**
500	**2282.9**	**4887.9**	**5048**	**5228**	**5303**	**6378**	**8013**

* 10 Kg. bags per acre

The table demonstrates that a farmer who harvested 50 bags will never make a profit at the prices quoted above; but if the harvest were 70 bags, then they must sell at M18.33 per bag to realize a profit of M131.00. That can happen if the potatoes are harvested when the prices are good; such as just after a prolonged drought or in late September through to January. On the other hand, if the produce is plentiful, e.g. 200 bags harvested; they can be sold at low prices like M6.87 per bag and the farmer can still make M221.00 profit.

Conclusion

The Machobane Farming System has proven to be effective in increasing the output of small and subsistence farmers in Lesotho. Through crop rotation, relay planting and intercropping, the system increases the intensity of land use, uses more of the growing season and reduces risk. It provides for lower production costs and opportunities for integrated pest management. The focus on potatoes as a field crop increases the potential for income generation. While the system has been developed specifically for the condition in Lesotho, many of the basic principles should be adaptable to other farming areas in Eastern and Southern Africa.

ENDNOTES

1. The author gratefully acknowledges the assistance of Dr. J.J. Machobane, 'Me Nora Klass, Axel and Elke Mayer, Thuso Green and the Mafeteng Development Project in the preparation of this paper.
2. Net Return = Gross Income - Total Costs (Fixed + Variable costs)

REFERENCES

Badamchian, Behjat. 1991: "Management of acid soils through liming", *Production Guidelines, RD- PG-15 Soils.* Maseru: Agricultural Information Services, Ministry of Agriculture, Cooperatives and Marketing.

Chandler, Z. 1963: *Handbook: Machobane Mass Agricultural College and Cooperative Union.* Unpublished.

Elwell, H. and Maas, A. 1995: *Natural Pest and Diseases Control*. Harare: Natural Farming Network.

Hogstrom, U. and Magnusson, L. 1991: *Soil Investigation in Maphutseng Valley, SW Lesotho for Location of Agricultural Experiment Plots*. Pilot Study for the FISC Project. Working Paper 162. Uppsala: Swedish University of Agricultural Sciences.

Government of Lesotho, 1991: *Production Guidelines, Volume II*. Maseru: Agricultural Research Division, Agricultural Information Services, Ministry of Agriculture.

Government of Lesotho, 1994 *Lesotho Situation Report 1976/77 - 1991/92*. Maseru: Department of Economics and Marketing, Ministry of Agriculture and Bureau of Statistics, Ministry of Planning.

Government of Lesotho 1997: *Vegetable Prices Weekly. 1991-1996:* Lesotho: Division of Marketing, Ministry of Agriculture, Cooperatives and Youth Affairs.

Lewis, L.J. 1959: *Machobane Cooperative Movement.* Community Development Bulletin, Vol. X, No. 4.

Litsinger, J. A., Moody, K. 1976: Integrated Pest Management in Multiple Cropping Systems. *Multiple Cropping.* ASA Special Publication 27. American Society of Agronomy, Crop Science of America, Soil Science of America.

Machobane, J.J.N. 1959: *The National Plight*. Unpublished.

Machobane, J.J. 1981: *Summary II: Machobane Farming System, Mant'sa Tlala.* Unpublished.

Machobane, J.J. 1988: *Training Basotho to produce more to feed their own people.* Unpublished.

Makhata, H., Marlowe, G., Campbell, J. and Jobo, T. 1991: *Production Guideline: Potato.* Agricultural Information Services, Ministry of Agriculture.

Ministry of Agriculture. 1979: *Soils of Lesotho.* Maseru: Conservation Division.

National University of Lesotho. 1990: Citation for Conferment of the Degree of Philosophy (Agriculture) (Honoris Causa).

Pfotenhauer, L. 1987: Profile: James Jacob Machobane. *Molepe, In flight Magazine of Lesotho Airways*, Vol 1, No 2.

Vukasin, H.L., Roos, L., Spicer, N. and Davies, M. 1995: *Production without Destruction*. Harare: Natural Farming Network.

Chapter 5

Mechanization Options for the Machobane Farming System

by Timothy E. Simalenga

Introduction

Lesotho has 300,000 hectares of arable land, on which about 270,000 tons of cereal, 23,000 tons of legumes and 70,000 tons of fruit can be produced. Theoretically, this is adequate to feed the entire Lesotho population and export some surplus (MOA, 1997). However, Lesotho's soil fertility and rangeland carrying capacity are in decline.

Dryland farming, which is the most widely practised in Lesotho, is hampered by several natural, human and socio-economic constraints. The major problem is adverse weather conditions – recurrent droughts, hail and frost. Dryland farming is characterized by declining production in basic grains and pulses and a serious degradation of the natural resource base, as evidenced by soil erosion and poor crop stand. This situation is further aggravated by communal use of land, overgrazing and encroaching settlements. Livestock feed on crop residue in the field, depriving rangelands of vegetative cover, thus making the soil more susceptible to soil erosion. About 15 tonnes of fertile soil per hectare is lost annually to erosion. Land classified as arable has been reduced from 13% in the 1960s to 9% in 1996.

One response to the land pressure and food security problems is the Machobane Farming System (MFS), which was developed by a Lesotho farmer. The system is mainly used by small landholders with approximately one acre of land. More than 80% of Lesotho's 2 million people live in rural areas; of these, 60% are farmers who depend on agriculture as their primary or secondary source of livelihood. Most farmers practice subsistence-level mixed farming of crops and livestock keeping. For these farmers, the Machobane system provides a foundation for self-reliant rural communities and for the eradication of poverty. The MFS is based on an intercropping strategy, with several crops grown simultaneously or in relay in the same field. Using this method, more than eight crops can be planted on the same piece of land and harvesting can be done three times a year. This makes the system unique, as it ensures that no one crop is destroyed and it ensures that the farmer has something to eat throughout the year. The system can be described as a low external input system, as the farmer depends on the resources of land, livestock, labour and household products, while the use of chemical fertilisers and pesticides is discouraged. The MFS offers an opportunity for integrated pest and soil management, as well as water conservation due to a cover crop being on the land at all times.

The system is promoted by the Machobane Agricultural Development Foundation (MADF) and has the support of the Ministry of Agriculture and the Rural Self-Help Development Association (RSDA), which started promoting the MFS in 1995. In the first year of the RSDA project, 75 farmers were registered. Within two years, RSDA had 300 registered farmers practising MFS. Currently, there are more than 2,400 registered farmers.

One of the main problems reported by farmers using the Machobane system is the "labour burden" and the drudgery of using a spade or hand hoe to open furrows, plant and turn the soil during crop rotation. The first intention of this study was to document the labour intensity of the Machobane system. The second purpose was to discuss various options with farmers and propose technologies and implements which can be used to improve labour efficiency.

Labour Intensity Under MFS

Data was gathered from key informants and through informal discussion with farmers. A questionnaire was designed and used to establish the labour input under MFS. This was then compared with labour input when using the traditional monocropping system. Farmers who completed the questionnaire were selected randomly from a list of registered farmers under RSDA.

Through analysis of sample daily work calendars (see Table 5.1), it is clear that farmers spend only two hours per day in the field and the maximum rows worked (opening furrow, manure application, planting, etc.) varies between two and three rows or lines per day. This is a very low labour/time input on the farm when compared to farmers from other countries in sub-Saharan Africa (Sylwander & Simalenga, 1997).

Table 5.1 Daily work calendar

Male Daily Calendar		Female Daily Calendar	
4.00 a.m.	Wake up	5.00 a.m.	Wake up, prepare food and children for school
5.00 a.m.	Feed fodder to cattle		
7.00 a.m.	Take cattle to communal veld and rest while cattle graze	7.00 a.m.	Clean house, prepare food for herdboys
11.00 a.m	Go to the field	10.00 a.m.	Go to the field
13.00	Look after the cattle	12.00	Prepare lunch
17.00	Back home and rest	14.00	Household chores (washing clothes/ironing), back to the garden
18.00	Supper	17.00	Back home and rest
19.00	Rest and socialization	18.00	Prepare supper
20.00	Go to bed	20.00	Go to bed

The daily work calendars indicated a division of labour according to gender. Livestock production activities, including herding livestock in distant communal grazing areas, are done by men and boys, while most agricultural field work is done by both men and women. However, most families in rural areas are headed by women, as the majority of men prefer to work in the mines or construction industries. In addition, almost all domestic work is done by women assisted by female children. Thus, while both males and females share the agricultural production tasks, women also perform other tasks which men do not normally perform and women end up with heavier workloads than men.

Figure 5.1 Comparison of labour input of MFS and traditional monocropping systems

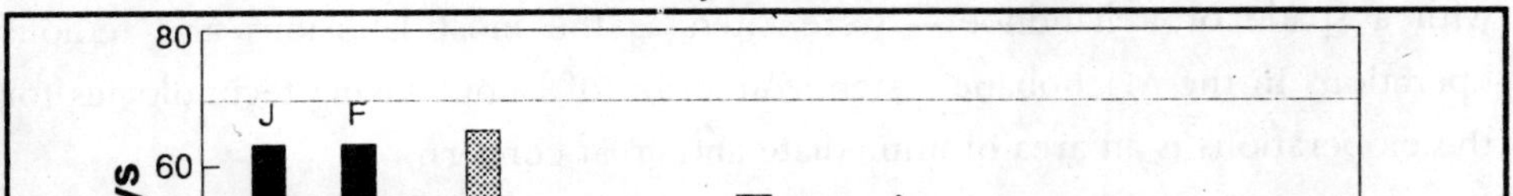

An attempt was made to compare the labour time spent on the farm in the traditional farming method of monocropping and the Machobane system. The labour input for each month was compared and significant differences were found only in the months of April, August and September. Otherwise, there is very little difference of labour input between the Machobane Farming System and the monocropping system (Figure 5.1). The real issue, therefore, is not the labour intensity of MFS as reported elsewhere, but the drudgery which needs to be reduced. Farmers identified eight operations involved in crop production under MFS. These included: furrow opening, manure/ash mixing and application, sowing, weeding, earthing of potatoes, harvesting, storage and marketing. Table 5.2 highlights the main operations, technologies/tools in use and their constraints.

Table 5.2 Current technologies used in the Machobane Farming System

Operation	Technology in Use	Remarks
tillage (1st year)	ox-plough	sometimes tractor can be used
furrow opening	spade or handhoe	row spacing depends on crop
planting	manually	all seeds
digging pits	spade or hand hoe	for planting potatoes
mixing manure/ash	spade	can be done on site or homestead
transport manure to the field	wheelbarrow	wash basins or tins also used
manure application	manually	efficient system required
weeding/soil turning	spade or hand hoe	efficient system required
earthing of potatoes	hand hoe	

Furrow opening, sowing, manure mixing and application – all done manually with a spade or a handhoe – were cited as the most laborious and tedious operations in the Machobane system. Adoption of labour-saving technologies for these operations is an area of immediate and great concern.

Mechanization Options

Mechanization in Lesotho has a long history. It is reported that farmers adopted the plough, learned to grow wheat and to ride oxen and horses during the 18th century. Animal traction has been a part of Lesotho's farming system for a long time (Bordet & Mhlanga, 1994). Cattle is the main draft power used on the farm (Table 5.3). Donkeys and horses seem to be widely distributed, but their use in agricultural production is limited to transport. Tractorization became significant in the 1970s, after independence. However, the performance of tractorization projects has been disappointing and none of them have shown a capacity for self-sustainability. A census carried out by the Ministry of Agriculture indicated that the population of tractors is about 1,300 and of these, only 65% were serviceable, while 25% needed minor or major repairs and 10% were beyond repair. Similar failures of tractor hire schemes and tractorization efforts by governments has been reported in other countries (Scheiman, Khalid & Simalenga, 1989; Simalenga, 1997).

Table 5.3 Use of draft animals, tractors and manual labour in Lesotho

	Tractors	Animals	Manual
ploughing	31%	69%	0%
planting	10	65	25
inter-row cultivation	<1	90	10
fertilization	5	0	95
spraying	<5	0	>95
harvesting	<5	0	>95
transport of harvest	20	80	0
threshing	20	40	40

Source: Bordet & Mhlanga, 1994

MECHANIZATION ALTERNATIVES

With the help of RSDA extension officers, farmers from ten different villages were assembled in two centres: HaSeetsi and Keketsi villages. During the meetings, farmers were asked to list their main problems and constraints with regard to MFS. Farmers were also given opportunities to suggest actions that might help solve the highlighted problems. To stimulate discussion regarding the alleviation of labour burden problems, a photographic display of tools and equipment which have been used successfully in other countries was presented. The main goal was to involve farmers in the planning and selection of mechanization options which would suit their environment. Farmers were later asked to identify and propose the tools or implements which they thought might be appropriate to their local environment.

Design and development of the hand-wheel push rippers

After group discussions and identification of possible tools which could be tried to alleviate the labour burden problems, two types of hand-wheel push rippers were designed. The first design had a single ripper tine (suitable for opening furrows between wheat rows). It was a combination of the wheel hoe (Figure 5.2) and the animal-drawn ripper tine (Figure 5.3).

Figure 5.2 A wheel hoe pushed between row crops

Figure 5.3 The animal-drawn ripper tine

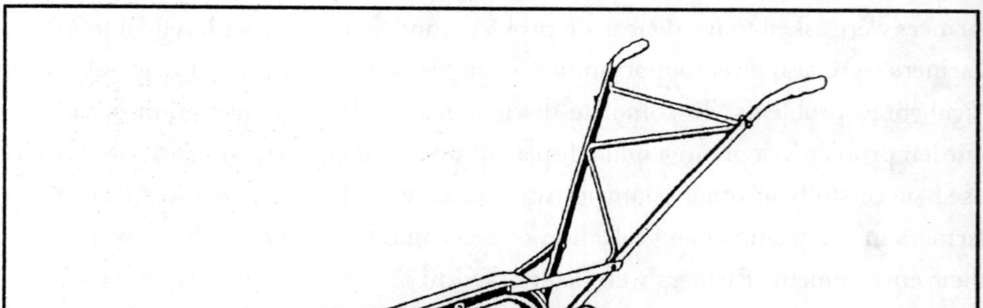

The single-tine hand-wheel push ripper can facilitate the furrow opening for planting maize, beans, sorghum, etc. between rows of other crops.

The other design had two-row ripper tines with planter attachments for planting wheat and peas (Figure 5.4).

Figure 5.4 The two-row hand-wheel push ripper with planter attachment

Two local manufacturers, Expo Hardware and Lesotho Steel, were asked to fabricate the two-row hand push rippers.

The animal-drawn ripper tine

In addition to the two-hand push instruments, an animal-drawn ripper tine was also made available. The animal-drawn ripper tine attachment was brought from South Africa, where it is being promoted by the Agricultural Research Council (ARC) in Pretoria and the Animal Traction Centre (ATC) of Fort Hare University. The ripper breaks up hard soil easily and makes it easy to plant seeds in a furrow at the same time. The attachment can easily be fitted to any "green" plough or ridger frame. A planter can also be fitted to the ripper.

On-farm Demonstrations

The two-hand wheel push rippers, together with the animal-drawn ripper tine, were then used for on-farm demonstrations. After the demonstration, farmers were asked to work with the implements in their fields, make comments regarding their performance and quality of work, and suggest any improvements that could be made. The following is the summary of the main comments made by farmers regarding the improvement of the implements.

- **The hand-wheel push ripper:** This ripper is a lightweight (about 9kg) implement which worked well in light soils. To improve its performance and to be able to work on hard soils, farmers proposed the following:
- The tine should be made wider and a bigger wheel (30cm diameter) should be used.
- To generate extra power when working in hard soils, a drawbar should be provided, to which a rope can be attached, allowing a second person to pull the implement from the front.
- The possibility of making the handle adjustable (for short and tall persons) was also suggested.

Figure 5.5 Opening a furrow between wheat rows using a hand-wheel push ripper

- **The two-row hand push ripper with planter attachment:** Women found it very difficult to work with this heavy implement (23kg). Farmers proposed that this implement should be pulled by a donkey, so a drawbar mechanism should be provided. Also, the tines should be made wider. However, to reduce the investment cost, it was suggested that the two-row tines with planter attachment be incorporated in the hand-wheel push ripper.

In this way, a farmer would buy only one handle/frame and two attachments. When planting wheat or peas, the two-row tines would be used. This could be removed and a single tine fitted for opening a furrow when planting maize and beans between the wheat or pea rows.

- **The animal-drawn ripper tine:** The demonstrated animal-drawn ripper tine raised a lot of interest among the farmers. This implement can be used to break the hard soil before planting potatoes. However, because the MFS row spacing of two meters is not wide enough for two oxen to walk comfortably, it was proposed that farmers should use a single ox or donkey with this implement. This will require single neck yokes and donkey harnesses. Farmers were eager to try using donkeys if the harnesses could be made available.

Recommendations and Follow-Up Issues

Following the study, a number of recommendations for improving labour efficiency under the Machobane Farming System and for making the system more comfortable for farmers were presented to the RSDA for consideration.

- **The hand-wheel push ripper:** At least five samples of the hand-wheel push ripper should be made and distributed to farmers in various villages, where they can be used for one season. This will generate interest and information regarding its performance and quality of work. Interested farmers should be encouraged to order the implement from a local supplier.
- **The animal-drawn ripper tine:** Training rural artisan/farmers and demonstrations on how to make single neck yokes and donkey harnesses will be required. To avoid re-inventing the wheel, proven designs from other countries can be used in the training and in making the yokes and harnesses.
- **Manure mixing:** Manure mixing and its application in the field was noted as another main time constraint. It is therefore proposed that a simple low-cost manure mixer and applicator/spreader be designed and introduced into the Machobane system.
- **Composting:** During the study, some farmers – especially those without cattle – complained about the unavailability of kraal manure and their neighbours' lack of interest in selling manure. It is strongly proposed that compost training be intensified.
- **Hand push weeder:** The possibility of using a hand push weeder should be

explored further. This type of implement can also help pulverize the soil and earth the potatoes.

- **Seed saving:** All farmers currently use hybrid seeds, which are made available and loaned by RSDA. This is not sustainable in the long run. The possibility of using local indigenous seeds, which can be stored by farmers, should be explored.
- **Networking and collaboration issues:** Networking and collaboration are important for exchanging information and stimulating linkages between farmers, local organizations and international organizations. At the farmer level, informal meetings should be organized and support for the establishment of farmers clubs should be given. These clubs will be a catalyst and a base to discuss and plan their activities, while empowering their decision-making processes. At the national level, linkages should be created amongst all organizations promoting the Machobane Farming System. This will allow for uniformity of the messages delivered to farmers and avoid duplication of extension manuals. An annual joint meeting of relevant organizations, such as MADF, MOA, National University of Lesotho, SWACAP etc., would help advance the MFS. At the international level, it is recommended that links be established between relevant organizations with similar interests. These may include SANAT (South Africa Network for Animal Traction), ATNESA (Animal Traction Network for Eastern and Southern Africa) and the FSR&E (Farming System Research and Extension) Association.
- **Publication of training and extension materials:** Farmers have very limited access to information and, as a result, many are not aware of alternative technologies and farm management issues. The MADF move to prepare training materials is highly commended. As part of farm management, farmers should be trained to plan their activities and keep farm records. Production and distribution of a calendar with pictorial sketches of activities to be done each month would be very helpful in assisting farmers to plan field activities and work on schedule for each operation.

REFERENCES

Bordet, D. and Mhlanga, M.L. 1994: *Agricultural Strategy Formulation for Lesotho. Maseru, FAO.*

Ministry of Agriculture (MOA). 1997: *Crop Production Policy Statement.* Maseru: Ministry of Agriculture, Department of Crops.

Moeller, O. 1997: *Farmers' tools: a description of agricultural tools used by farmers in the East and Southern African region*. Harare: FAO/Farmesa.

Scheiman, D., Khalid, A. and Simalenga, T.E. 1989: *The feasibility of tractor mechanization in Tanga region, Tanzania.* Dar-es Salaam: GTZ/GoT.

Simalenga, T.E. (ed.). 1994: *Participatory Research and Development of Agricultural Engineering Technologies.* Proceedings of on-farm trials regional workshop. Harare: AGROTEC.

Simalenga, T.E. (ed.). 1997: *Farm Mechanization and Strategy Formulation in East and Southern Africa.* Proceedings of a regional workshop. Harare: FAO/FARMESA.

Simalenga, T.E. and Joubert, A.B.D. (eds.). 1997: *Animal Traction in South Africa. Today and Tomorrow.* Proceedings of SANAT workshop held March, 1996. Alice: SANAT.

Sylwander, L. and Simalenga, T.E. 1997: *Gender Issues in Animal Traction; A handbook.* Harare: ATNESA.

CHAPTER 6

MANURE MANAGEMENT OPTIONS FOR INCREASING CROP PRODUCTION IN THE COMMUNAL AREAS OF ZIMBABWE

by Herbert K. Murwira and Jean K. Nzuma

INTRODUCTION

The majority of smallholder and communal farmers in Zimbabwe occupy areas that consist mainly of sands or sandveld soils that are inherently infertile. These areas are extremely marginal for crop production. Nitrogen (N), Phosphorus (P) and Sulphur (S) are critically deficient in these soils (Mashiringwani, 1983; Grant, 1981). The inherent low soil fertility is worsened by widespread soil degradation, in the form of declining soil fertility, loss of organic matter, breakdown of soil structure and erosion (Elwell & Stocking, 1988).

Livestock manure has been used by communal and smallholder farmers to supply nutrients to the crops. When applied to the soil, manure is often a valuable fertilizer and soil amendment. Manure has been reported to improve water-holding capacity, improve soil structure and moderate soil temperatures. In addition, yearly applications of manure have been shown to increase the fertility of sandveld soils by progressively increasing the cation exchange capacity, exchangeable bases and pH (Grant, 1967). However, significant losses of nutrients during storage and handling of manure by leaching and volatilization can greatly reduce the effectiveness of manure as a source of nutrients.

The use of crop residues is also fairly widespread in communal farming areas. Crop residues remain as roots or above-ground stover in the fields at the end of the dry season. Farmers also directly incorporate residues. Observations in Mangwende indicate that at the onset of the season, some farmers prefer to burn off remaining crop residues, which they claim impede tillage operations. Burning also helps to destroy pests, pathogens and weed seed. The ash produced during burning can enhance plant growth by reducing soil acidity and reducing nitrogen immobilisation, as crop residues have high C:N ratios. However, burning results in a loss of N, which then has to be replaced with expensive inorganic fertiliser.

Alternative and socially acceptable methods of managing crop residues have to be developed. One possible method is the age-old, but probably forgotten, art of composting. Only about 12% of farmers were reported to be using compost in Mangwende in 1985 (Shumba, 1985). Use of compost tends to be popular amongst poorer households, where it is used for both maize and vegetable gardens (Carter & Murwira, 1995; Shumba, 1985). A concerted effort is therefore required to promote the use of compost as an alternative technology for managing soil organic matter. It is imperative to carry out studies on composting technologies and methods to manipulate its quality.

Manure management in kraals and utilisation of crop residues are intricately linked in many ways. Crop residues are fed to cattle in the kraals, resulting in an inadvertent mixture of manure and crop residues. Farmers also directly apply crop residues and grasses to manure in the kraals for use as bedding materials or to increase the bulk of manure. The diversity of these crop residue management strategies requires a holistic approach.

The beneficial effects of manure and crop residues on soil fertility, and in particular their role as sources of plant nutrients, depend on the outcome of the biological processes of decomposition. These processes determine the rate of release of carbon and nutrients as well as the equilibrium between retention and loss from soil (Swift, Heal & Anderson, 1979). The amount of nitrogen in the final manure is critical if the manure is to be valuable for plant growth. Despite low nutritive value of maize stover fed to cattle during the dry season (Topps & Manson, 1967), it is a valuable source of nutrients – the cattle excrete 80% to 90% of the nitrogen, phosphorus and potassium they consume (Mentis, 1981). The return of nutrients from crop residues fed to cattle is potentially high (Grant , 1970).

Losses of nutrients, especially nitrogen, through leaching and volatilization during manure storage and handling significantly reduce its effectiveness and quality as a nutrient source. The nitrogen content of manures from communal areas range from 0.5% to 1.8%, but they are frequently less than 1.2% (Tanner & Mugwira, 1984; Murwira, Swift & Frost, 1995). The low N content is probably due to poor-quality feed and losses during storage. Studies reported in the international literature show that nutrient losses during storage and handling can be substantial. Kirchman (1985) found that between 8% and 40% of manure N is lost during storage. Lauer, Bouldin and Klausner (1976) measured losses of between 61% and 99% of ammonium nitrogen, while Beauchamp, Kidd & Thurtell (1978) obtained losses of between 56% and 60% ammonium nitrogen from anaerobically digested sludge.

The use of crop residues during the decomposition of manure can reduce nitrogen losses during storage and handling. Addition of stover (with initial C/N ratios above 30) to cattle manure significantly reduced N losses. Adsorbents such as peat and zeolite are also effective in reducing ammonia losses during decomposition. Peat and zeolite placed in the spent air-stream reduced ammonia losses by 59% and 16% respectively (Witter & Kirchman, 1989). In pot experiments conducted by Kirchman (1985), uptake of manure N from aerobically treated manures was lower than from anaerobically decomposed manure with the same C/N ratio. Cecchi, Mata-Alvarez and Vertstraete (1992) concluded that organic fractions could, by anaerobic composting, be converted to compost of excellent quality. Kirchman (1985) reported that nitrogen concentration increased during the first six months of storage; after six months, the nitrogen concentration declined.

There are few documented studies on improving manure quality in Zimbabwe's communal areas. In a comparison of quality traits (sand, nitrogen, phosphorus and potassium content) of manure from different types of kraal floors, Nyathi, Khombe and Dube (1992) found that manure from concrete floors had less sand than those with earth floors. Manure from concave slopes had a higher proportion of sand than manure gathered from sloped floors. Similar trends in the amounts of N, P and K in manure from the kraals were reported in the study.

There is still a paucity of information on manipulating biological processes to improve the quality of manure during decomposition. This study aims at improving traditional soil fertility management methods by developing manure storage and handling systems that minimize nutrient losses and enhance manure quality.

Storage conditions affect the quality of manure and the subsequent nutrient release. Mugwira (1985) noted an increased nutrient uptake by the second crop for aerobically decomposed manure, but insignificant P uptake. Anaerobically decomposed manure had a positive nitrogen effect on crops in the short term, while aerobically decomposed and dried manure showed a net immobilisation of N in the first eight weeks after application (Murwira *et al.*, 1995). This suggests that anaerobic storage of manure might improve N availability. Studies on decomposition of communal area cattle manures showed that they mineralise less than 10% of the total in the first season of application (Murwira & Kirchmann, 1993). This suggests that the high C:N ratio could be limiting the rate of N release. Therefore, improving manure storage and handling conditions and adding inorganic N should both reduce the C:N ratio and promote faster decomposition by increasing the N available to microbes and hastening the release of manure into inorganic form.

Manure and compost are only two of a range of nutrient sources communal area farmers use in what amounts to a complex, sophisticated strategy of soil nutrient management. Farmers also use leaf litter, termitaria soil and fertiliser. But manure accounts for 57% of the mass and 87% of the total N of locally available inputs used by farmers. When the total N budget is taken into account, inorganic fertiliser supplies 64% and manure 32% of the total N farmers use (Murwira *et al.*, 1995). Manure makes a significant contribution to the overall N budget, even on farms with high access to inorganic fertilisers. It is interesting to note that almost all farmers who use manure and compost apply them to the maize crop (Shumba, 1985). Given the ever-decreasing quality and quantity of manure available to farmers, we consider it a viable strategy to increase the value of inputs already available to the farmers. For example, increasing the quality of manure N by 0.1% will provide an extra 10kg N at application rates of 10t/ha – or the equivalent of about 30kg of ammonium nitrate.

The strategy for improving the use of manure and crop residues involves three interventions:

1. Improve the quantity and quality of the manure produced by supplying the animals with more food and food of better quality.
2. Improve the quality of manure and crop residues through more efficient composting techniques. This would require interventions in collection, processing and reducing nutrient losses.
3. Improve on-field management of organic materials. This involves application rates, application methods, efficient distribution of a limited resource, inorganic nutrient supplementation and so on.

The work reported in this study focuses on the second and third interventions. The study addressed several key questions:

- What are farmer-appropriate composting techniques that enhance nutrient availability while reducing losses?

- What are the residual effects, and how should composted products be managed on the farm?
- Can the composted materials be stabilised to make their nutrient composition more uniform, so that the rates or combinations of the organic and inorganic fertiisers can be tailored to fertiliser recommendations based on soil tests?
- Are there any advantages to composting manure/crop residues and applying the materials after two years, as compared to the usual one-year cycle?

The Study

The study was conducted in Mangwende communal lands. In Phase I, a multi-disciplinary team made a detailed survey of the use of crop residues and manure and the storage and handling techniques of cattle manure. Farmers were asked how they manage their manures and residues, why they do not adopt recommended interventions, which interventions they could adopt, and what resources they have. Participatory rural appraisals, informal interviews, wealth ranking and bio-resource flow modeling techniques were used to develop a research agenda that was sensitive to the needs of the target communities. Bio-resource flow models developed by farmers were used as a basis for developing crop residue and manure management strategies. The practical interventions that were proposed following the survey – and that farmers considered undertaking – were tried in on-farm or on-station experiments.

In Phase II, manure storage experiments were conducted. Outdoor experiments were carried out on cow dung storage, using mixes of locally available crop residues. Compost piles of manure and crop residue mixes were allowed to decompose under aerobic and anaerobic conditions. Weighed amounts of fresh, well-mixed manures were placed in pallet boxes and allowed to decompose for between four and six months. Two sets of on-farm trials were implemented; comparisons of pit and surface storage of manure and comparisons of improved composting techniques (using crop residues) with current farmers' practice. In all the experiments, the manures were fully characterised in terms of lignin content, N, P, K, Ca, Mg and micronutrient content. Determinations of nutrient amounts and dry matter losses in composts were made at different times during the experiments. Processes involved in nutrient losses during decomposition were studied in incubation experiments in order to fine-tune the strategies in the manure storage experiments described above. Field experiments with the various manure by-products of the storage experiments were established to derive correlations of their nutrient availability and uptake by the plants. These experiments tested the manures produced with different storage techniques for their efficacy on maize growth. Trials were established to test manures from both farmers' practices and the improved practices. A total of 11 trials were established in the 97/98 growing season.

In Phase III, there was more emphasis on the on-farm work. Parallel farmer-managed trials were initiated in year 2 to screen and adapt technologies in the study areas. Simplified trial designs (versions of Phase II experiments) were formulated with the farmers, and at least 14 farmer-managed trials were established. Compost-making demonstrations were given in order to popularise the technologies. Extension played an increasing role during these demonstrations.

Results and Discussion

Analysis of storage and handling practices

The survey established that current management techniques are poor and lead to large losses of potential nutrients. Eight distinct practices were identified in the storage and handling of manure. These were: selection of the kraal site, kraal feeding (addition of litter to the kraal), heaping manure, deep stall (manure kept in the kraal until mature), trapping seepage in pits or ridges, placing stones on upper slopes of the kraal or constructing walls around the kraal and rotation of kraals.

The heaping and deep-stall practices were distinguished as storage options in winter (Table 6.1). The heaping practice was followed by 86% of the farmers in the sample. Four percent of these farmers incorporated fertiliser additives in their heaps.

Table 6.1 Winter manure storage and handling practices

Storage conditions	Length of composting	% of farmers
Heaping	1 week to 6 months	82
Heaping plus fertilizer additives	2 to 3 months	4
Deep stall plus addition of litter	4 to 6 months	4
Deep stall without addition of litter	4 to 6 months	10

The duration of heaping ranged from a few weeks to six months, with a mean duration of about three months (Table 6.2). A majority of farmers heaped between the months of August and October.

Table 6.2 Duration of heaping and frequency distribution amongst farmers.

Duration of heaping	Frequency amongst farmers (%)
1 to 3 weeks	15
1 to 2 months	27
3 to 4 months	27
5 to 6 months	8
> 6 months	9

Of the 14% of farmers that practised the deep stall method, 10% did not add crop residues in their kraals.

Ameliorative practices to counter run-off losses included rotation of kraals (8%), trapping seepage (28%), constructing thick walls or placing stones on the upper slopes of the kraal (52%). The proportion of farmers following each practice is indicated in parentheses. Only 12 per cent of the farmers practiced no ameliorative technique.

Kraal feeding was perceived as a preliminary treatment in the handling process. Combinations of organic materials were used as bedding materials. However, there were variations in the times at which residues were added to the kraal. Forty percent of farmers added residues only in the wet season. The kraal site was perceived to play a role in either minimising or promoting nutrient losses. Of interest were kraals that were erected on anthills – the quality of manure from these sites was perceived to be high.

Farmers were able to establish what they perceived to be the reasons for nutrient losses in manure and their solutions to the problem. Figure 6.1 summarises these views.

Figure 6.1 Farmer perceptions of causes and solutions for nutrient losses from manure

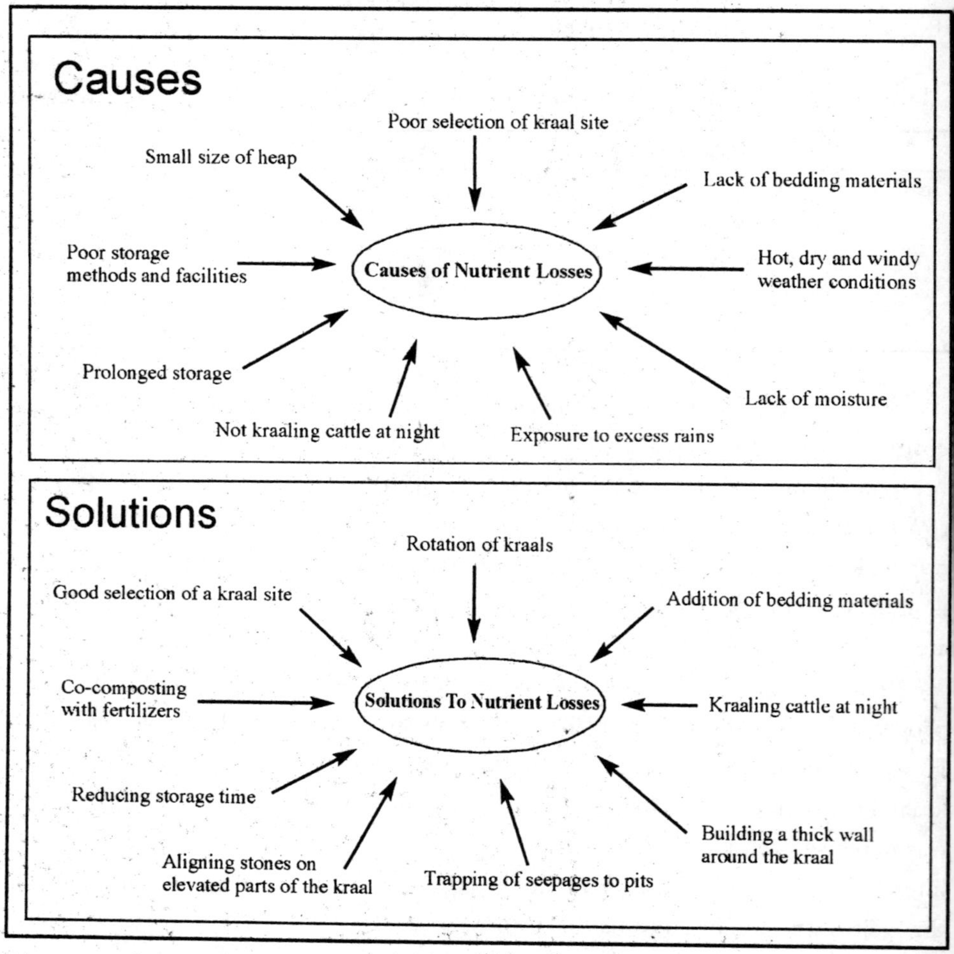

The nutrient status of manure samples collected during the survey was generally low. In terms of their N content, 67% of the samples were deficient, while 30% and 3% were marginal and rich respectively. This trend applied also to the rest of the nutrients (Table 6.3).

Table 6.3 Quality of manure and frequency distribution amongst farmers

Chemical composition of manure	Frequency	Distribution amongst farmers (%)
N content	*< 0.5 - 0.90	67
	1.00 - 1.20	30
	> 2.25	3
P content	*< 0.15 - 0.2	51
	0.25 - 0.30	43
	> 3.0	6
K content	*< 0.5 - 1.00	65
	1.00 - 2.00	39
	> 2.50	6

* Deficient range

Several factors were found to have contributed to the low quality of manure collected. These included the conditions under which manure was stored and handled, ambient temperature, moisture levels and exposure to the environment. Although the conventional heaping practice saves labour and produces manure in time for application, it exposes manure to direct sunlight, allows for aeration which triggers NH_3 volatilisation and results in aerobically dried and decomposed manure low in nutrient contents. Inefficient use of crop residue in the kraal was observed. Variations in times at which residues were used indicate there are periods when nutrients are not totally conserved, particularly at the critical point of excretion. High losses from urine were reported when there is low or no use of adsorbents (Kirchmann & Witter, 1988).

The points of weakness identified in the current management systems provided a basis for formulating a research agenda that was sensitive to the context. Manure quality was improved by manipulating biological processes during storage and handling. Laboratory incubations and field experiments were conducted to assess the efficacy of the solutions that emerged from the initial survey.

Effectiveness of straw as an ammonia absorbent in cattle kraals

Incubation studies were made using dynamic techniques to determine the quantities and time course of ammonia loss from freshly excreted dung and urine. Ammonia volatilisation during manure handling can cause serious nitrogen losses, which reduce its agronomic value. It is important to use efficient management techniques to minimise such losses. Ammonia loss can be reduced to low levels by incorporating manure into soil, but this is not effective when the greatest potential losses occur before the manure is incorporated into soil (Murwira, 1995). Losses are especially large when the manure is mixed with urine and/or the decomposition process is aerobic.

In such cases, mixing a nitrogen-poor material such as straw with the manure and urine can minimise nitrogen losses. The efficiency of using straw to reduce N loss was examined in the laboratory. Different amounts of straw were used to determine the optimum quantity required to reduce ammonia losses from manure. Zero, 1, 2, 3, and 4g of straw were mixed with 25g batches of cow dung and placed in dreschel bottles. Airflow scrubbed free of ammonia, at a pressure of 0.02 bars, was blown through the dreschel bottles, and the ammonia produced was trapped in 2% boric acid. In other experiments, ammonia losses were also measured from urine, manure, and from mixtures of straw with either urine or manure.

The results from these experiments showed that straw was an effective ammonia adsorbent. A proportion of three parts straw to 25 parts manure was found to be the most effective in reducing losses. Addition of straw reduced losses by up to 85% in cow dung, and by 50% in cow dung and urine mixtures (Fig.6.2).

Figure 6.2 Effects of straw in reducing losses of N

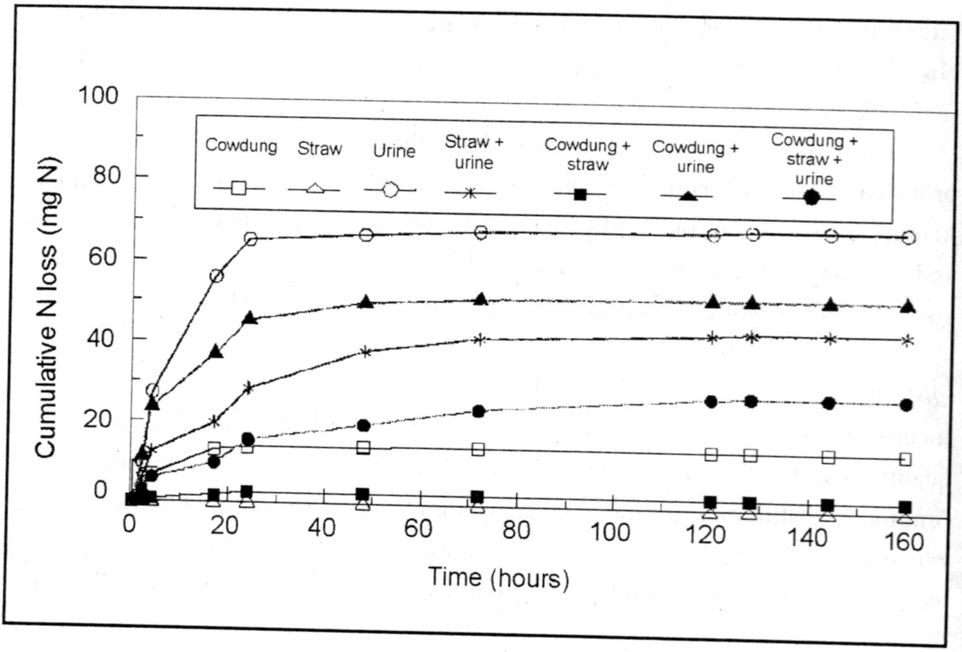

While the results from the laboratory study show that it is feasible to mix straw and dung to minimise nitrogen losses, further research is needed, especially on a field scale, to determine the efficacy of the technique. In a field study done in Mangwende Communal Area, Zimbabwe, losses were measured in cattle kraals with or without crop residues used as bedding. Losses from cattle kraals with crop residues were 80% less than in kraals without crop residues.

Effect of storage method, crop residue incorporation and duration of storage

Pit storage of manure (anaerobic composts) combined with use of crop residues in both summer and winter storage was identified as a promising technology, which was evaluated against the conventional heaping technique (aerobic composts). Results from these experiments showed that treatment effect, length of composting and treatment by month interactions were significant ($P<0.001$).

Significantly larger amounts of total N were measured in anaerobic (pit stored) manure compost than in aerobic (heaped) manure composts. Nitrogen concentration of manure decomposing in pits was almost three times higher than that of manure decomposing in heaps, particularly when residues were not added. The total N concentration of manure in pits increased gradually with storage time. The low nitrogen concentration in the heaps can be attributed to loss of N through ammonia volatilisation, which occurred during storage.

The highest NH_4^-N concentrations, amounting to 225ppm and 380ppm for April and July storage systems respectively, were measured in pits without maize residues. Pits incorporating maize residues recorded 280ppm and 155ppm for July and April systems respectively. This implies that incorporating crop residue has a strong effect on manure quality. In contrast, lower concentrations of 50ppm were recorded for manure decomposing under the heaps for both periods. The use of straw in heaped manure composts showed no significant effect on NH_4^-N concentrations. In general, ammoniac-N levels increased with a longer composting period. However, ammonium-N levels were higher in pit-stored manure than in heaped manure, The highest NH_4^-N concentrations found in pits were due to anaerobic conditions that prevailed during decomposition and resulted in accumulation of NH_4^-N. The higher moisture levels maintained by the pit system resulted in a greater capacity of the manure to store N in the ammoniac form.

The opposite scenario was observed for the nitrate-N, which was higher in heaps than in pits. The results also show increased nitrate concentrations with addition of straw. The use of straw improved aeration and the general conditions suitable for nitrification. There was also an increase in soluble-N with time particularly during the third and fourth month of storage.

Little or no N losses from pits were recorded. Highest NH_3 losses from pits amounted to 7ppm and were from systems where crop residues were not incorporated. The low losses were possibly due to low circulation of air and the closed system that prevented drying and NH_3 volatilisation. Higher NH_3 losses amounting to 35ppm were measured during aerobic decomposition in heaps where crop residues were not incorporated. The greatest losses occurred between August and October, the last two months of composting. This scenario was probably triggered by exposure of heaps to the environment which induced aerobic conditions, drying and faster rate of decomposition and high mineralisation of N.

Significantly lower ($P<0.001$) NH_3 losses of between 5ppm and 10ppm were measured in anaerobic manure composts than in aerobic manure composts. In pit storage systems, the pH did not rise above 7, which explains why little NH_3 was lost. The pH under anaerobic systems remained slightly acidic or neutral throughout the decomposition period. In contrast, the pH values recorded for aerobic manure composts were above 8. This is in accordance with the pH dependent equilibrium between ammonium-ions and ammonia gas in hydrous solution. Further, the initial NH_4^-N in heaped manures, which might have been derived from the hydrolysis of urea and uric acids in the dung, could have resulted in a temporarily high pH, thus creating conditions for NH_3 volatilisation. However, the addition of maize straw in aerobic decomposition heaps significantly reduced NH_3 losses throughout these systems.

There was a marked reduction in moisture content in heaps as decomposition progressed, compared to pits, which lost their moisture gradually. The higher moisture loss in heaps probably explains the high losses of NH_3. Addition of residues resulted in a faster loss of moisture.

Effect of crop residue entrapment of urine in the kraal.

In an experiment to determine the effect of crop residue entrapment of urine in the kraal, high NH_3 losses were found in kraals where residues were not used (Table 6.4). The largest losses were recorded at midday. Significant reduction in NH_3 losses from manure was obtained by adding straw in the kraal. Cereal crop residues absorb urine and have high C/N ration which leads to temporary immobilisation of N. Crop residues are therefore a way of conserving nutrients during handling.

Table 6.4 NH_3 losses (ppm) measured during a 4 week period from kraals with and without crop residue incorporation.

Kraal Number	Time of taking readings		
	8 am	12 pm	3 pm
	A. Kraals Without Crop Residue Additions*		
1	15.50	37.59	13.10
2	13.08	47.17	15.00
3	7.92	28.00	13.70
	B. Kraals With Crop Residue Additions		
4	2.04	4.65	1.13
5	3.42	3.55	1.17
6	2.25	3.04	1.30

* Figures given are mean values recorded over a month (n=4).

The effect of pit-stored and heaped manure on nutrient efficiency and maize yield.

Trials were carried out to determine the efficacy of the different manure management options that were developed during the study. Manure from the storage experiments was used. These experiments were conducted from November 1997 until the end of the rainy season in April 1998. Results shown in Fig.6.3 indicate significantly higher yields from pit-stored manure than heaped manure at the same application rate.

Conclusion

The sets of results presented clearly demonstrate that pit storage of manure combined with use of crop residues can minimise NH_3 losses by volatilization and enhance manure quality. The challenge is to evaluate nutrient use efficiency of pit-stored and heaped manure and determine management options for applying manure on sandy soils. There is also need to determine the mineralization pattern of manures from pits and heaps, as the fertiliser value of manure depends, among other factors, on the rate of mineralization of nutrients to the inorganic form. These studies are also important because efficient management of manure-N

Figure 6.3 Effects of manure storage treatment on maize yields

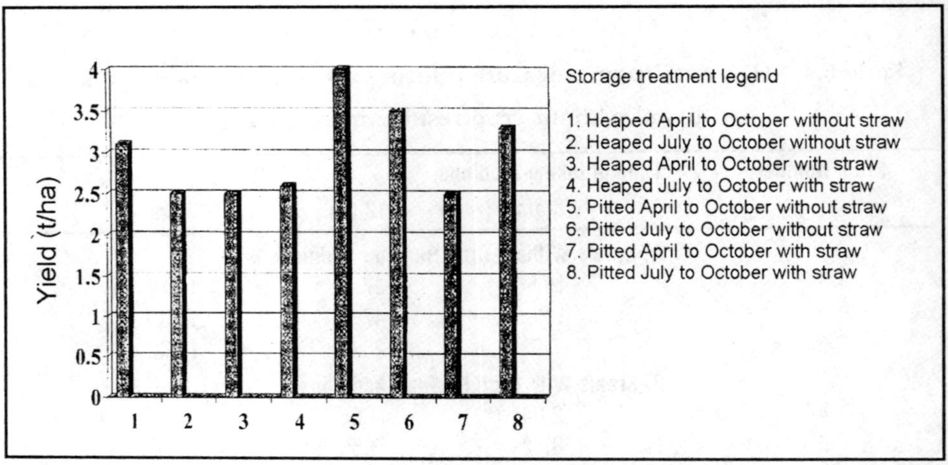

requires knowledge of the pattern of mineralization and N transformations. These studies will further determine processes that lead to NH_3 losses in soil following application of manure.

References

Beauchamp, E.G, Kidd G.E and Thurtell G. 1978: Ammonia volatilization from sewage sludge applied in the field. *Journal of Environmental Quality* 7: 141-146

Carter, S. and Murwira, H.K. 1995: Spatial variability in soil fertility management and crop response in Mutoko Communal Area, Zimbabwe. *AMBIO*, 24 (2):77-84.

Cecchi, F., Mata-Alvarez, J. and Vertstraete, W. 1992: Memorandum of the International Symposium on Anaerobic Digestion of Solid Waste held in Venice. In: *Waste Management International* pp595-600

Elwell, H.A. and Stocking, M.A. 1988: Loss of nutrients by sheet erosion as a major hidden farming cost. *The Zimbabwe Science News* 22 (7/8) pp79-82

Grant, P.M. 1967: The fertility of sandveld soil under continuous cultivation. *The Rhodesia, Zambia and Malawi Journal of Agricultural Research*. 5 pp71-79 and pp117-128

Grant, P.M. 1970: Changes in the fertility of para-ferralitic sands under continuous cropping. Paper presented at the Soil Science Society of South Africa conference, April, 1970.

Grant, P.M. 1981: The fertilization of sandy soils in peasant agriculture. *Zimbabwe Agricultural Journal* 78 (5) pp169-175

Kirchman, H. 1985: Losses, plant uptake and utilisation of manure nitrogen during a production cycle. *Acta Agriculturae Scandinavia. Supplementum* 24

Lauer, D.A., Bouldin, D.R. and Klausner S.D. 1976: Ammonia volatilization from dairy manure spread on the soil surface. *Journal of Environmental Quality* 5: 134-141

Mashiringwani, N.A. 1983: The present nutrient status of the soils in the communal farming areas of Zimbabwe. *Zimbabwe Agricultural Journal* 80 (2): 73-75

Mentis, M.T. 1981: Direct effects of animals on pasture and veld. In N.M. Tainton, ed., *Veld and Pasture Management in South Africa*. Natal: Shutter and Shooter in association with University of Natal. pp299-321.

Mugwira, L.M. 1985: Effects of supplementing communal area manure with lime and fertilizers on plant growth and nutrient uptake. *Zimbabwe Agricultural Journal* 82 (5):153-159

Murwira, H.K. 1995: Ammonia losses from Zimbabwean cattle manures before and after application to soil. *Tropical Agriculture* (Trinidad) 72: 269-274

Murwira, H.K and Kirchmann, H. 1993: Comparison of carbon and nitrogen mineralisation of cattle manures subjected to different treatments, in Zimbabwean and Swedish soils. pp189-198. In R. Merkx and J. Mulongoy, eds., *The Dynamics of Soil Organic Matter in Relation to the Sustainability of Tropical Agriculture*, Leuven, Belgium: John Wiley & Sons.

Murwira, H.K., Swift, M.J. and Frost P.G.H. 1995: Animal-Soil interactions: Manure As A Key Resource in Sustainable Agriculture. pp 131-148. In J.M. Powell, *et al.* (eds.) *Livestock and sustainable nutrient cycling in mixed farming systems of sub-Saharan Africa* Ethiopia: ILCA.

Nyathi, P., Khombe, C.T. and Dube, I.A. 1992: The effects of kraaling and stover supplement during the season on body weights and manure production of Mashona steers in Zimbabwe. *African Livestock Research* 1: 18-23

Shumba, E.M. 1985: On-farm research priorities resulting from a diagnosis of the farming systems in Mangwende, a high potential area in Zimbabwe. *Zimbabwe Agricultural Journal*, Special report No.5.

Swift, M.J., Heal, O.W. and Anderson, J.M. . 1979: Decomposition in terrestrial ecosystems. *Studies in Ecology Vol 5*. London: Blackwell Scientific Publications.

Tanner, P.D. and Mugwira, L.M. 1984: Effectiveness of communal area manure as a source of nutrients for young maize plants. *Zimbabwe Agricultural Journal* 81 (1): 31-35

Topps, J.J. and Manson, J.L. 1967: The maize stover after early harvesting of grain. *Rhodesia, Zambia and Malawi Journal of Agricultural Research*. 5: 191-192

Witter, E. and Kirchman, H. 1989: Peat, zeolite and basalt as adsorbents of ammoniacal nitrogen during manure decomposition. *Plant and Soil* 115: 43-52.

CHAPTER 7

CATTLE MANURES IN ZIMBABWE: METHODS AND MINERALS FOR QUALITY IMPROVEMENT

by Peter van Straaten

INTRODUCTION

Many soils in Zimbabwe's marginal cropping areas are strongly degraded and deficient in organic matter (Grant, 1970). This is especially true of the soils in communal areas where over 70% of Zimbabwe's rural population live (Mehretu, 1994). Soil degradation has increased dramatically over the last decades, affecting millions of hectares of land. Not only is the nutrient status of the soil declining, soil erosion is also a serious problem. Over 20% of Zimbabwe's farmland has been classified as moderately to very severely affected by soil erosion (Whitlow, 1988). About 91% of the eroded cropland is located in the densely populated communal areas, compared to only 8% in the large-scale commercial farmland (Rukuni, 1994).

The main organic fertilizer used by smallholder farmers is cattle manure, which is a mixture of cattle faeces and urine. Nearly all of the readily available nitrogen (in the form of ammoniacal nitrogen) comes from the urine. Nitrogen from the faeces is largely unavailable for the crops and it will only become available after organic matter breakdown.

The application of cattle manure will increase soil nutrients, increase organic matter, increase biological activities, improve the soil structure, increase the infiltration of water, increase water holding capacity, decrease soil surface crusting, decrease soil bulk density reducing resistance to root penetration, and reduce risks to soil erosion by wind and water. Farmers know of the physicochemical and biological soil improving qualities of manure. Cattle manure is regarded as the fertilizer of choice for most farmers in the communal areas because it bears the least risk for crop failure. However, the nutrient status of the cattle manure used by most farmers in the communal areas is poor, specifically in regard to nitrogen and phosphorus. Because of poor soil quality, the vegetation is also nutrient poor. Animals browsing on nutrient-poor vegetation produce nutrient-poor manure. Nevertheless, the total nitrogen output of cattle is significant.

There are an estimated 4,400,000 cattle in Zimbabwe (1992 estimate, Tawonezvi, Sukume, Zitsanza & Rukuni, 1994). These produce a manure nitrogen content of approximately 1.0% in communal areas per year. The nitrogen content of manures from commercial feedlots is commonly higher due to high-value feed (1.5 - 1.9% N). On a mass balance basis, this means, cattle in Zimbabwe produce approximately 44,000 - 75,000 t N annually. In terms of nitrogen values alone, the manure is worth approximately US$ 26 - 45 million. At least another 40,000 t of nitrogen are estimated to be lost in or on the way to the grazing areas and as a result of volatilization and leaching. The total annual 'N-production' by cattle in Zimbabwe is estimated at 80,000 - 120,000 t N.

The only nitrogen fertilizer industry in Zimbabwe, Sable Chemical Industries Ltd., produced approximately 56,000 t N in 1992/93 (Sable Chemical Industries Ltd., 1994). However, the ammonia processing facilities of Sable Chemical Industries Ltd. use the electrolysis process for manufacturing ammonia, a process which consumes an enormous amount of electricity (approximately 8,000 kWh per tonne of ammonia).

This study contributes to the understanding of the current manure management system in communal areas of Zimbabwe focusing on new ways of nutrient resource conservation and improvement. Smallholder farmers can utilize indigenous nutrient resources more efficiently and sustainably by reducing nitrogen and phosphorus losses from manures. Improved manure handling can make a significant contribution to the nutrient budget of the country as a whole and to the preservation of soil quality.

The Study

Detailed descriptions of specific manure management techniques from smallholder farmers in Sub-Saharan African countries are relatively rare (Probert, Okalebo & Jones, 1995; Kwakye, 1980; Musa, 1974). The aim of the study was to assess the quality of local manures and to test locally available inorganic materials for their effectiveness in reducing the losses of nitrogen and increase the phosphorus content of animal manures in livestock enclosures. The study sought:

- to quantify the nutrient composition of cattle manures in some areas of Buhera District
- to quantify the nutrient losses from manures
- to assess the capacity of organic and inorganic additives, specifically inorganic mineral additives and mine wastes for improving manure quality

In addition, the study sought to collect independent data on the two components of manure, urine and faeces. While there are many detailed studies of manures and manure handling and storage systems in some of the more developed countries in the North, there are few specific and quantitative data available from the two distinctly different components of cattle manures (Brady 1990; Hall, 1920; Salter & Schollenberger, 1938).

Figure 7.1 The study area

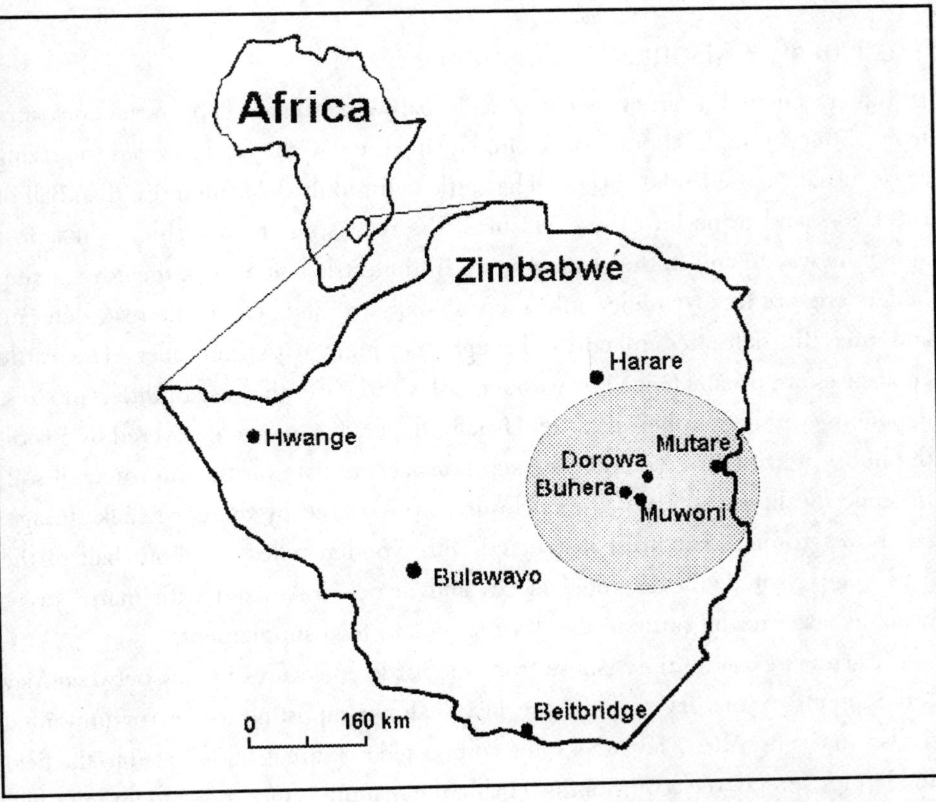

Two short field studies were carried out in the Buhera District, Manicaland Province, Zimbabwe (Fig. 7.1) by the author and Mr. A. Mbwera of the Institute of

Mining Research (IMR), University of Zimbabwe. Discussions on strategies and logistics were also held with Dr. R.T.C Fernandes of IMR. The field visits in April/May and October 1996 gave effect to the principle that farmers should play a more prominent role in adaptive agriculture research and development. The researchers worked with many farmer families: the Chinodya family at Murambinda, the Mparadzi family at Shawa, the Nhambura family at Nerutanga, the Makwere and Mbwera families and more than 20 other farmers from the Buhera District. Extension workers from AGRITEX were involved in the collection of data, in preliminary analyses of manures in the field, and in discussions on the current status of manures and on manure handling and storage improvements. For comparative purposes, visits were also made to a small-scale dairy farm in the Buhera District and a commercial dairy farm near Harare.

The Current Manure Management System

In Buhera District, as in many parts of Zimbabwe, cattle are kept in the enclosure from about 5 p.m. to about 7 or 8 a.m. During the day, the cattle are led to grazing areas, usually close to the villages. The cattle commonly defecate more than half of the faeces and urine into the enclosures, the rest is not recoverable, as it is lost along the way to and in the grazing areas. In the cattle enclosures, the cows, oxen, heifers etc. are free roaming, only a few animals are tied. The cattle rest, defecate and mix the defecated manures through trampling with their feet. The cattle enclosures are usually fenced by wooden poles and wire; they are of different sizes, depending on the number of cattle. Usually, the enclosure is not covered by a roof, the floors are not covered and the excretions accumulate on the unprotected soil. In some localities, however, the enclosures are covered by the corn stalk storage, an above-ground structure constructed with wooden poles. In about half of the cattle pens visited, the excreted faeces and urines are mixed with maize straw, which is given to the cattle in the dry season as a feed supplement.

Farmers excavate the manure from the cattle enclosures usually between May and September (the dry season), and heap it for composting to destroy unwanted seeds and weeds. After a few weeks, the composted manure is transported to the field and left for several weeks or months. Finally, the manure is incorporated into the soil, shortly before the onset of the rainy season, usually in October. By the time the manure is incorporated into the soil, it is usually completely dry and low in nutrients.

Quantity Estimates

Khombe, Dube and Nyathi (1992) measured the amount of manure collected in research cattle pens using yearling Mashonaland steers with an average liveweight of 170 kg. The annual recoverable amount of bedding-free manure was determined as 424.4 kg (dry weight basis). In another study, physical collection and direct weighing of excrements released by an animal over a period of one day were carried out near Muwoni and Murambinda, resulting in a preliminary estimate of the excreted amounts. On average, the 150-200 kg live weight local Shona cows excreted 10-12 kg faeces (wet weight basis) per day, and 4-6 litres of urine per day. The result is approximately 800 to 1000 kg (dry weight basis) recoverable manure (including decomposed bedding) in the cattle pen every year (van Straaten, field notes, 1995). These field data are in good correlation to estimates of the American Society of Agricultural Engineers (ASAE, 1981). According to ASAE, the amount of faeces and urine are a function of the body weight of cattle. Taking 150 kg and 200 kg liveweight as estimated average weight of cattle in this part of Zimbabwe, the amount of faeces and urine would amount to 12.3 to 16.4 kg/day/head. An average North American cow with 1000 lbs (454 kg) live weight produces 37.2 kg "raw waste" per day (ASAE, 1981).

Quality Measurements

The manure data collected thus far in Zimbabwe have not differentiated between contributions of the urine and faeces fractions respectively. The current study attempted to do this. Daily visits to cattle pens were made during various periods of the day to collect fresh samples for chemical analysis. Ten samples of fresh cattle faeces and 125 samples of manures were collected and analyzed on their nutrient content. Forty urine samples were collected in plastic buckets "in free air" and analyzed in the field for their pH. Thirty-four urine samples were collected in sterile vacuum test tubes and refrigerated until analyses for N,P,K,NH_4. The samples were analysed either at the Soil Productivity Laboratory of the Department of Research and Specialists Services in Marondera, Zimbabwe, or at the University of Guelph (Analytical Services Laboratory, Department of Land Resource Science). Classical analytical techniques were used: Kjeldahl N-analysis, P colorimetric tests, K, Ca, Mg and Zn, Cu using Atomic adsorption techniques, NO_3 and NH_4 analyses using the Technicon Auto Analyzer (TRAACS). The data compiled in the course of this survey contribute to the data base of chemical analyses of cattle manures in

Zimbabwe and complement data published by Mugwira (1984), by Mugwira and Mukurumbira (1986), and by Tanner and Mugwira (1984).

Analysis

The chemical composition of 38 manures from cattle pens in the Buhera District and two from a dairy farm close to Harare were analyzed. The statistical means of N, P, K, Ca, Mg, Zn, Cu concentrations from these samples are given in Table 7.1. The pH of the manures were all alkaline, they ranged from 8.68 to 10.11 with a mean value of 9.58.

Table 7.1 Chemical analyses of 38 untreated cattle manures in Buhera District, collected in April/May 1996

Total	N	P	K	Ca	Mg	Zn	Cu
	%			μ g/g			
Mean	1.112	2566	13504	8653	3087	43.8	11.3
S.D.	0.377	1765	6817	4150	1742	23.0	8.9

The chemical analyses of 17 urines are presented in Table 7.2.

Table 7.2 Chemical analyses of 17 untreated cattle urine samples from Buhera District, collected April/May 1996

Total	N (%)	K (%)	NH4 (mg/kg)
Mean	0.4635	1.384	1470.82
S.D.	0.104	0.387	430.68

Based on these data on manure quantity and quality, it is possible to calculate the nutrient production under the current manure management system. At an average annual manure production rate of 800 kg and 1,000 kg (dry weight base) per cow respectively, and at an average composition of N = 1.1%, P = 0.25%, K = 1.35% , the nutrient quantity per head of cattle per year in the Buhera District communal area will be approximately:

9 to 11	kg N,
2 to 2.5	kg P,
11 to 13.5	kg K.

Given that the amount of readily available nutrients from cattle manure is generally only approximately 1/2 (Brady, 1990) or less, it will be necessary to apply between 20 to 30 tonnes of manure per hectare in order to meet the nutrient requirements of a crop like maize. The effect of cattle manure applications will last for two to three years. The readily available nutrients will be released in the first year. In subsequent years, more nutrients will become available as the manure decomposes.

Nutrient Variability

The chemical composition of cattle manures varies widely from place to place. The variability is ascribed to factors such as age and condition of animal, kind and amount of feed consumed, nature and amount of litter, construction aspects of the cattle pen and the storage of the manures (Brady, 1990; Kwakye, 1980; Hall, 1920, Salter & Schollenberger, 1938). In the study area, a considerable variability in manure nutrient content was observed. There were differences in the vertical distribution of the cattle manure, and there were differences in the surface layer within one cattle pen. In general, higher nitrogen values were found in samples with higher moisture contents. Higher nitrogen contents were found in the more compacted, moist (probably partially anaerobic) dark grey to black manures below the surface. In the light and fluffy surface manures, significant differences were observed within the cattle pen, some higher nitrogen values certainly attributable to urine patches. Urine patches were easily detected in the cattle pen by means of ammonia gas detection using the Gastec detection system.

During the April/May survey two unusual manure samples were collected from the cattle pen of Mr. and Mrs. Mparadzi, at Shawa. As shown in Table 7.3, the concentrations of phosphorus, calcium and magnesium values in their sample were distinctly different from the average manures. The faeces sample from this pen was also high in nutrients.

Table 7.3 Comparison between average cattle manure composition and Mparadzi's cattle manure composition

Sample	N %	P %	K %	Ca %	Mg %	Zn ppm	Cu ppm
Average cattle manure	1.11	0.26	1.35	0.86	0.31	43.8	11.3
Mparadzi's cattle manure	1.01	0.73	2.98	1.64	0.88	57.0	13.5

It was suspected that the higher nutrient concentrations in these manures were a result of different forages. The Mparadzi cattle pen is located near to the Shawa igneous complex, a rock type relatively rich in magnesium, calcium and phosphorus. During the October survey, this association was studied more closely. Manure samples were collected from cattle pens located in two distinctly different geological and soil environments: in granitic areas with relatively poor nutrient status and in the Shawa area, which is characterized by high nutrient status of rocks and soils. The results of the October survey are presented in Table 7.4.

Table 7.4 Chemical analyses of untreated cattle manures collected outside the Shawa area and in the Shawa area in October 1996

Total Sample	N %	P	K	Ca g/g	Mg	Zn
Mean outside Shawa	0.79	1453	8338	6496	2423	33.1
Mean from Shawa	1.00	5032	16337	12043	5500	43.4

The data demonstrate that the manures collected from areas underlain by the P-, Ca- and Mg-rich alkaline complex of Shawa had significantly higher P, Mg, K, Ca concentrations as compared to manures from cattle browsing in granitic areas. The ratio of nutrient in manures from the Shawa area to the nutrient concentration of manure in granitic areas is as follows:

P: 3.46
Mg: 2.27
K: 1.96
Ca: 1.85
N: 1.26

The highest contrast is in the P content, followed by Mg. The nitrogen concentrations in manures over granite were only slightly lower than in the Shawa area.

A comparison between manures from the communal areas and manures from commercial feedlots has been made by Mugwira and Mkurumbira (1986). The data collected during our field visit include four samples (out of 80) from commercial dairy farms. The nitrogen and phosphorus values were clearly higher in the commercial dairy manures (by up to 50%). The strongest contrast was observed in the zinc and copper content of the manures, which were more than five times higher in manures from commercial feedlots than in the communal manures. The higher values are a clear reflection of the different feeds, the commercial dairy cattle having been given a concentrate, enriched in the trace elements zinc and copper, among others (Table 7.5).

Table 7.5 Comparison between mean cattle manure composition of communal Buhera District and 2 commercial dairy farms outside the Buhera District

Sample	N %	P %	K %	Ca %	Mg %	Zn ppm	Cu ppm
Mean of 38 cattle manure compositions from Buhera District	1.11	0.26	1.35	0.86	0.31	43.8	11.3
Mean of four cattle manure compositions from commercial dairy farms outside Buhera District	1.54	0.61	0.83	No Data	No Data	238.0	76.5

NUTRIENT LOSSES

Manure handling systems are known for their high nutrient losses, especially nitrogen. It is reported that more than 50% of nitrogen in cattle manures is lost due to inadequate manure management techniques, volatilization and leaching (Beauchamp, Kidd & Thurtell, 1982; Beauchamp & Burton, 1985; Brady, 1990; Dewes, 1995; Dewes, Schmitt, Valentin & Ahrens, 1990; Muck & Richards, 1983; Safley, Westerman & Barker, 1986; Termeer & Warman, 1993).

Volatilization

The major pathway of nitrogen loss is through the volatilization of ammonia gas. Ammonia gas (NH_3) develops upon conversion of urea from excreted urine through the enzyme ureas and from drying of manures containing ammonium (NH_4^+). The rate of ammonia loss is primarily a function of the concentration of ammonia in the manure, the concentration of ammonia in the surrounding air and the rate of exchange across the liquid-gas interface. The equilibrium between ammonia gas and ammonium is influenced by the pH and by temperature (Salter & Schollenberger, 1938).

The amount of volatilization of nitrogen from urine is difficult to measure under field conditions. In the field study, ammonia emissions were measured in the cattle pen and over manure heaps by means of a Gastec portable gas detector using ammonia tubes for testing. The distribution of ammonia in a cattle pen varied significantly, both over the enclosed pen area and vertically within the accumulated cattle manure in the pen. The background values of ammonia over manure (measured 5 cm above the material to be tested) were in the range of 3-5 ppm ammonia. The highest ammonia gas values were measured over freshly excreted urine, with ammonia values reaching up to 30-40 ppm. The highest ammonia values were measured in freshly dug sections of a cattle pen. In the zone of the dark, compacted cattle manure, usually 5-10 cm below the surface, ammonia contents reached 50 ppm.

An indication of N losses over time is provided by comparing chemical composition, specifically N concentrations of manures in the same area during different periods of time. In the Buhera District, the 38 manure samples collected in May showed a mean N concentration of 1.12%. Samples taken from the same area in October, after being excavated, mixed and exposed to dry hot weather conditions for one to two months had mean N concentrations of only 0.84% (52 samples). This corresponds to a gross nitrogen loss of 32%. Since little if any rain falls during this period, it is assumed that most of the N losses are caused by volatilization.

Leaching

The cattle enclosures in large parts of the Buhera District are underlain by coarse to medium grained sandy soils, a porous medium with a high permeability and thus a high probability of leaching nutrients from cattle pens.

Several pits were dug in cattle pens to sample underlying soils in order to study possible leach effects. Data from five of these profile sections indicate that various nutrients are leaching from the manures in the cattle pens into the underlying sandy soils. Nitrate values in the sands below the pens ranged from 34.8 to 108.6 mg/kg. In general, at a depth of 25-30 cm below the contact with the manures, the nitrate values were in the range of 23-81 mg/kg. The ammonium values in the sand under the cattle pens were in the range of 3-4 mg/kg, with the exception of the section under the cattle pen of Chinodya family, where 23.5 mg/kg were measured directly at the cattle manure/sand contact. It can be concluded that considerable amounts of nitrogen, especially nitrate, leach out of the cattle pen. The possibility that the nitrate leaching from these sources may contaminate local drinking water should be investigated.

Potassium and phosphate have also been leached from the cattle manures. Elevated values of potassium were observed in the sandy soils directly under the cattle pens, reaching 5610 mg/kg (average content in sands = 500-1000 mg/kg). Phosphorus, commonly regarded as very immobile, has also leached into the underling soils, probably as part of the organic matter. A few centimetres below the manure/sand contact, it reached up to 350 mg/kg (average 5-35 mg/kg).

An Experiment on Reducing Nutrient Losses

A recognized way of reducing nitrogen losses through volatilization is to reduce the pH of the manures or, more specifically, the main urea carrier in manure, urine (Dewes, 1995; Frost, Stevens & Laughlin, 1990; MacKenzie & Tomar, 1987; Pain Thompson, de la Lande Cremer & ten Holte, 1987; Stevens, Laughlin & Frost, 1992; Termeer & Warman, 1993). A second approach is through adsorption of ammonium on minerals like smectites, vermiculites or zeolites. These minerals are all characterized by their large surface areas, high cation exchange capacities (CEC) and ammonium selectivity. Vermiculites and smectites have high ammonium adsorption, making the nitrogen compounds less prone to volatilization. It is known that adsorption of ammonium occurs within hours after contact and basic laboratory studies indicate that the rate of ammonium adsorption and subsequent release from these minerals and from mineral-organic matter complexes are of potential use. But the mechanism of retaining ammonium in the manures by adsorption of has not yet been studied in detail.

The study included a simulation experiment to measure changes in nutrient contents of manures exposed to the ambient climatic environment and to find practical and relatively uncomplicated ways of reducing nitrogen losses and increasing phosphorus levels using these two methods. A homogenized sample of the cattle manure from the Nhambura family was used. This is a manure for which many analytical data already exist. The cattle manure was collected at the end of April 1996 and thoroughly mixed. Several control samples were taken, in order to see the variability of this material. The other component was a composite urine sample, collected "in-air" from nine cows in the Muwoni area. The pH of the composite urine sample was 8.26. The homogenized sample of cattle manure and the composite cattle urine sample were mixed in a ratio 2.2 to 1 (manure to urine) and kept in plastic dishes exposed to the sun on the roof of a farm building in Muwoni, Buhera District.

The inorganic phase added the following minerals and mineral products to the composite manure and urine samples:

- **Phospho-gypsum:** Phospho-gypsum is a waste material produced during the production of phosphoric acid and triple superphosphate. At the phosphate processing plant of Zimphos in Masasa/Harare, several hundred thousand tonnes are stored on "waste" piles. The management of the processing plant are actively looking for new markets for this waste product (Mr. H. Ruzvidzo, ZIMPHOS Ltd., pers. comm. 23.4.1996).
- **Single superphosphate (SSP):** SSP is currently being produced by Zimphos Ltd. from local phosphate (from Dorowa Minerals Ltd) and iron sulphide (pyrite) resources (from Iron Duke Mines Ltd).
- **Pelletized phosphate blends (PPB):** Pelletized phosphate blends (PPB) have been developed by the Zimbabwe-Canada agrogeology project, supported by the International Development Research Centre (IDRC), the University of Guelph and the Zimbabwean Government (van Straaten & Fernandes, 1995). It is a 50:50 blend of triple superphosphate and Dorowa rock phosphate.
- **Dorowa tailings:** Twelve samples from the very fine-grained "slimes" of the Dorowa phosphate mine tailings were collected by drilling. The old tailings piles of Dorowa are estimated to contain approximately 8 million tonnes of these fine-grained materials, disposed as waste (Mr. R. Chitata, Dorowa Minerals Ltd., pers. comm. 24.4.1996).
- **Shawa vermiculites:** The Shawa vermiculite mine processes 50-100 t of vermiculite per day. Around the processing plant, there are large waste heaps made up of the undersize vermiculite fraction.

Eighteen triplicate manure samples were collected on April 27, 1996, before and after treatments with these mineral additives. These were analyzed on May 3, 1996.

The pH of the urines collected during this field visit were all alkaline, ranging from pH 7.44 to 8.88, with a statistical mean of 8.14. In the experiment, the phospho-gypsum, single superphosphate and pelletized phosphate blends were all acidic and served to lower the pH of the manure. The vermiculite from Shawa with a pH 7.9 and the Dorowa tailings materials with a pH of 8.3 did not lower the pH of the manure (Table 7.7). These latter two adjuncts served as adsorption media.

Table 7.7 P_2O_5, CO_2, and pH of additives to cattle manure used in experiment

Sample	ID P_2O_5	CO_2	pH
Phospho-gypsum	0.53	0.09	2.3
SSP	17.55	0.12	1.7
PPB	40.38	0.39	2.3
Dorowa tailings			8.3
Shawa vermiculite			7.9

Results of experiment

The treated and untreated manures were sampled before and after a one-week exposure and analyzed on their nutrient contents.

- **Nitrogen:** The highest relative increase in total nitrogen after one week exposure was achieved by mixing pelletized phosphate blends with the manures. The nitrogen losses using these acidifying phosphate additives were lower than without additions. The results of this short-term experiment are in line with work carried out by MacKenzie and Tomar (1987) and reported from older reports on control of N-losses (Fink, 1982).
- **Phosphorus:** The best results in terms of increased phosphorus levels were achieved by using locally produced phosphates. The P levels of the manure treated with pelletized phosphate blends (PPB), at a rate of 12.5 g/kg of manure, increased from a mean value of 2,566 mg/kg to 3,860 mg/kg and more. The PPB showed the highest P increase because they represented the most concentrated P sources. The treatment with 6 g of single superphosphate (SSP) per kg of manure increased the P levels only slightly. Old reports describe the practice of using superphosphates to increase the P value of manures and reduce the nitrogen losses (Hall, 1920; Salter & Schollenberger, 1938; Fink, 1982; Brady, 1990). These methods are no longer practised in North America and Europe because of different manure handling systems and the general nature of manures with higher P levels, which are of environmental concern in those countries.

  Higher than average Ca contents were also found in manure samples treated with phospho-gypsum additives.
- **Magnesium:** The analytical data show that the Dorowa tailings materials and

the vermiculites fines from Shawa did not perform significantly better than the control in terms of improving the nitrogen and phosphorus status of the manure. However, both the treatments with vermiculite and Dorowa tailings increased the amount of available Mg content by a factor of 1.5 to 2.

Conclusion

The aim of the study was to assess the quality of local manures and to test locally available inorganic materials for their effectiveness in reducing nitrogen losses and increasing the phosphorus content of animal manures in livestock enclosures. It can be concluded that, under current manure management practices, the loss of nutrients through volatilization and leaching are high and that the use of mineral-amended manures, particularly pelletized phosphate blends, can contribute to nitrogen retention and increased phosphorus levels. Thus, the use of adjuncts can contribute to the reduction of soil degradation and an increase in soil productivity.

It will be necessary to continue this manure management research and participatory technology development with the farmers. There is a need for participatory research in regard to practical modifications of the cattle pens for increased urine retention and nutrient conservation. Tying up cattle in rows overnight can provide specific sites for faeces and urine deposition where inorganic amendments could be added. Construction of a furrow and pit storage system outside the cattle pen for urine collection and nitrogen conservation will also help to concentrate manure. Construction of clay-lined floors could prevent leaching and reduce the mixing of manure with underlying sands. The clays for the cattle pen floorings should be locally available. One of the resources that could be used is the fine-grained tailings materials from Dorowa. Roofs can be built over cattle pens, using the current stalk storage constructions. Outside the pens, stalk mulching of heaped manure can reduce the drying rate. The incorporation of manures into soils shortly after excavation, possibly in October, could also increase N retention.

Practical work with the contact farmers should include the building of two or three pilot cattle pens with local resources. The improved pen design will be complemented with pilot testing of the locally available pelletized phosphate blends and with the evaluation of manure qualities and crop yield trials.

The research conducted in this project was not only directed towards the improvement of cattle manures in the Buhera District. The use of locally available organic and inorganic resources is part of a broader strategy to find technical solutions to make more efficient and more ecologically balanced use of Southern Africa's natural resources, in cooperation with and for the benefit of the rural poor across the region.

REFERENCES:

American Society of Agricultural Engineers (ASAE). 1981: Manure production and characteristics. *Agricultural Engineers Yearbook 1981*, p. 442.

Beauchamp, E.G., Kidd, G.E. and Thurtell, G. 1982: Ammonia volatilization from liquid dairy cattle manure in the field. *Can. J. Soil Sci.*, 62, 11-19.

Beauchamp, E.G. and Burton, D.L. 1985: *Ammonia losses from manures.* Factsheet 85-071. Toronto: Ontario Ministry of Agriculture. and Food.

Brady, N.C. 1990: *The Nature and Properties of Soils*, 10th ed. New York, N.Y.: Macmillan Publishing

Carter, S.E. and Murwira, H.K. 1995: Spatial variability in soil fertility management and crop response in Mutoko Communal Area, Zimbabwe. *Ambio* 24, 77-84.

Dewes, T. 1995: Nitrogen losses from manure heaps. In *Nitrogen leaching in ecological agriculture*, pp. 309-317.

Dewes, T., Schmitt, L., Valentin, U. and Ahrens, E. 1990: Nitrogen losses during the storage of liquid livestock manures. *Biol. Wastes* 31, 241-250.

Dhliwayo, D.K.C. and Mandiringana, O.T. 1994: *Greenhouse evaluation of compacted phosphate blends.* Internal Report. Marondera: Department of Research and Specialist Services, Soil Productivity Research Laboratories.

Fink, A. 1982: *Fertilizers and Fertilization.* Verlag Chemie. Weinheim-Deerfield Beach-Basel.

Frost, J.P., Stevens, R.J. and Laughlin, R.J. 1990: Effect of separation and acidification of cattle slurry on ammonia volatilization and on the efficiency of slurry nitrogen for herbage production. *J. Agric. Sci.*, 115, 49-56.

Grant, P.M. 1970: Restoration of productivity of depleted sands. *Rhodesia Agric. Journal*, 67, 131- 137.

Grant, P.M. 1981: The fertilization of sandy soils in peasant agriculture. *Zimbabwe Agric. J.*, 78, 169-175.

Hall, A.D. 1920: *Fertilizers and manures*. London: John Murray.

Jacobsen, S.T. 1988: Storage of animal slurries by composting with straw. In *Storing,*

Handling and Spreading of Manure and Municipal Waste, Vol. 1. Uppsala: Technical Section of C.I.G.R.

Khombe, C.T., Dube, I.A., and Nyathi, P. 1992: The effects of kraaling and stover supplementation during the dry season on body weights and manure production of Mashona steers in Zimbabwe. *African Livestock Research* (1992) 1, 18-23.

Kwakye, P.K. 1980: The effects of dung storage and its nutrient (NPK) content and crop yield in the north east savannah zone of Ghana. In *FAO Soils Bulletin 43: Organic Recycling*, pp. 282-288.

Lauer, D.A., Bouldin, D.R. and Klausner, S.D. 1976: Ammonia volatilization from dairy manure spread on the soil surface. *J. Environ. Qual.*, 5, 134-141.

MacKenzie, A.F. and Tomar, J.S. 1987: Effect of monocalcium phosphate monohydrate and aeration on nitrogen retention by liquid hog manure. *Can. J. Soil Sci.*, 67, 687-692.

Mehretu, A. 1994: Social poverty profile of communal areas. In M. Rukuni and C.K. Eicher (eds.), *Zimbabwe's agricultural revolution*, pp. 56-69. University of Zimbabwe Publications.

Muck, R.E. and Richards, B.K. 1983: Losses of manurial nitrogen in free-stall barns. *Agric. Wastes*, 7, 65-79.

Mugwira, L.M. 1984: Relative effectiveness of fertilizer and communal area manures as plant nutrient sources. *Zimbabwe Agric. J.*, 81, 85-90.

Mugwira, L.M. and Mukurumbira, L.L. 1986: Nutrient supplying power of different groups of manure from the communal areas and commercial feedlots. *Zimbabwe Agric. J.*, 83, 25- 29.

Musa, M.M. 1974: A method for conservation of cattle manure. In *FAO Soils Bulletin 27: Organic materials as fertilizers*, pp. 89-95.

Nyamapfene, K. 1991: *The soils of Zimbabwe*. Harare: Nehanda Publ.

Pain, B.F., Thompson, R.B., de la Lande Cremer, L.C.N. and ten Holte, L. 1987: The use of additives in livestock slurries to improve their flow properties, conserve nitrogen and reduce odours. In V.D. Meer, R.J. Unwin, T.A. Van Dijk and G.C. Ennik (eds.), *Animal Manure on Grassland and Fodder Crops*, 229-246.

Probert, M.E., Okalebo, J.R. and Jones, R.K. 1995: The use of manure on smallholder's farms in semi-arid Kenya. *Expl. Agric.*, (1995) 31, 371-381.

Rodel, M.G.W., Hopley, J.D.H. and Boultwood, J.N. 1980: Effects of applied nitrogen, kraal compost and maize stover on yields of maize grown on a poor granite soil. *Zimb. Agric. J.*, 77, 229- 232.

Rukuni, M. 1994: *Report of the Commission of Inquiry into Appropriate Agricultural Tenure Systems.* Report for His Excellency The President of the Republic of Zimbabwe, 2 volumes.

Sable Chemical Industries Ltd. 1994: *The nitrogenous fertilizer industry in Zimbabwe, a review of Sable Chemical Industries Ltd. of Kwekwe.* Internal Report.

Safley, L.M., Westerman, P.W. and Barker, J.C. 1986: Fresh dairy manure characteristics and barnlot nutrient losses. *Agric. Wastes*, 17, 203-215.

Salter, R.M. and Schollenberger, C.J. 1938: Farm manure. *In Soils and Men, Yearbook of Agriculture: 1938*, pp. 445-461. Washington, D.C.: US Department of Agriculture.

Singh, R.D. and Dhar, N.R. 1986: Effect of organic matter without and with rock phosphate on crop yield, quality and soil characteristics. *Indian J. Agric. Sci.*, 56, 539-543.

Stevens, R.J., Laughlin, R.J. and Frost, J.P. 1992: Effects of separation, dilution, washing and acidification on ammonia volatilization from surface-applied cattle slurry. *J. Agric. Sci.*, 119, 383-389.

Tanner, P.D. and Mugwira, L.M. 1984: Effectiveness of communal area manures as sources of nutrients for young maize plants. *Zimb. Agric. J.*, 81, 31-35.

Tawonezvi, H.P.R., Sukume, C., Zitsanza, N. and Rukuni, M. 1994: *Beef sub-sector in Zimbabwe.* Harare: World Bank.

Termeer, W.C. and Warman, P.R. 1993: Use of mineral amendments to reduce ammonia losses from dairy-cattle and chicken-manure slurries. *Bioresource Techn.*, 44, 217-222.

van Straaten, P. and Fernandes, T.R.C. 1995: Agrogeology in Eastern and Southern Africa: a survey with particular reference to developments in phosphate utilization in Zimbabwe. In T.G. Blenkinsop and P.L. Tromp (eds.), *Sub-Saharan Economic Geology*, pp. 103-118. Geol. Soc. Zimbabwe Spec. Publ. 3. Netherlands: Balkema Publishers.

Whitlow, J.R. 1988: *Land degradation in Zimbabwe: a geographical study.* Harare: Department of Geography, Univ. of Zimbabwe.

Wild, A. (ed.). 1988: *Russell's soil conditions and plant growth*, eleventh edition. UK: Longman Scientific & Technical.

CHAPTER 8

MANURE EXTRACT IN VEGETABLE GARDENS: ON-FARM PARTICIPATORY RESEARCH IN MALAWI

by Rachel Bezner Kerr

INTRODUCTION

Manure extract is promoted as an alternative fertilizer for resource-poor farmers in countries such as Zimbabwe, Kenya, the Philippines and the United States. The method involves placing manure in a porous sack and placing the sack in a container of water (3 parts water to 1 part manure) for approximately two weeks. The sack is removed and the manure extract is used to fertilize crops[1]. It is not known, however, whether manure extract provides adequate nutrient sources or whether it is appropriate for the conditions of resource-poor farms. There has been no published scientific research. One undergraduate study in California tested manure extract as a fertilizer and found that it provided adequate levels of the major nutrients (N, P & K), but lacked sufficient levels of Ca, Fe, Mn and Cu (Price, 1985).

In 1995, the Development Department of the Church of Central Africa Presbyterian in Malawi began a training program for smallholder farmers. The programme demonstrated various methods of improving soil fertility and soil and water conservation. During these courses, the staff learned that some farmers were very interested in the use of manure extract. Several farmers requested more specific application rates and timing information from the Department. In cooperation with the Development Department, on-farm experiments were conducted to compare manure extract to other fertility management options now used by farmers in their vegetable gardens. These experiments made it possible to determine the effectiveness of manure extract under local conditions.

Soil Fertility and Nutrient Sources in Malawi

Soil surveys in Malawi indicate a wide variability of nutrient status (Young & Brown, 1962) and previous soil studies have found that most Malawian soils are deficient in nitrogen (Mughogho, 1989; Vlek, 1993; Saka , Bunderson, Itimu, Phombeya, & Mbekeani, 1994). As in other parts of Eastern and Southern Africa, resource-poor farmers in Malawi use a variety of soil nutrient sources, including crop residues, manures and fertilizers (Mughogho & Ngwira, 1996; Kasembe, Semoka & Samki, 1983). Fertilizer use in the smallholder sector increased rapidly from the 1960s through the 1980s. However, prices rose rapidly in the 1980s and fertilizer use dropped. Fertilizer use in Malawi remains low compared to global averages (Mughogho, 1989).

Most of the manure research in Malawi has focused on '*khola*' manure: cattle and/or goat manure in outdoor livestock pens. Open *kholas*, the most common type of livestock enclosure, are exposed to leaching and gaseous losses of nitrogen, making the manure less valuable as a fertilizer. There is little nutrient data about goat, pig or slaughterhouse manure in Malawi. Most of the trials conducted have compared a known weight of "farmyard manure" to a known weight of fertilizer in a given area. Many of the studies did not consider different nutrients separately or report the level of nutrients in the manure. Some trials report a low nitrogen response, while others report a significant response. Yield responses to manure have also varied markedly due to the variability in nutrient contents, with some trials indicating significant response and others suggesting no response (Mughogho & Ngwira, 1996).

Manure Nutrients

Chemical composition of manures varies widely due to various factors, such as age, condition of animals, type and amount of feed consumed, nature and amount of litter and storage conditions (MacLean, Miller & Robinson, 1983). The quality of livestock manure used by resource-poor farmers is often low because their livestock forage on soils with low nutrient contents and do not receive feed additives (Murwira & Kirschmann, 1993). Studies from nearby countries such as Tanzania, Zimbabwe and Kenya indicate that resource-poor farm manure has low levels of N, P, K and Ca (Murwira, Swift & Frost, 1995; Probert, Okalebo & Jones, 1995; Kasembe, Semoka & Samki, 1983). Manure handling systems can lose up to

50% of available nitrogen through volatilization, poor management techniques and leaching during storing, handling and application (van Dijk & Sturm, 1983; Barry, 1996).

Nitrogen losses in manure storage

Ammonia volatilization is the main pathway of nitrogen loss during storage. Ammonia (NH_3) is a gas at normal atmospheric temperatures and pressures; it can quickly volatilize to the atmosphere. Factors which influence volatilization include pH, temperature and solution ammonium (NH_4^+) concentration. At higher pH and temperature values, there is a much greater potential for volatilization (Freney, Simpson & Denmead, 1983). Wind speed close to the surface can also increase NH_3 losses. Volatilization losses begin almost immediately following excretion and are highest initially, decreasing markedly over time. Under aerobic manure storage, alkaline conditions develop which leads to rapid NH_3 volatilization, followed by slow NH_3 release. Anaerobic manure storage may lead to the production of volatile fatty acids (VFA) which decrease the pH, reducing NH_3 volatilization (Kirschmann & Witter, 1989). However, other studies have found that anaerobic manure storage induces the formation of ammonium carbonate, which increases pH (Kirschmann & Witter, 1992). A rising pH would increase volatilization rates. The concentration of water soluble nitrogen (NH_4^-N) and carbon/nitrogen (C:N) ratios tend to increase during anaerobic storage, which may decrease volatilization rates (Kirschmann & Witter, 1989).

In a comparison of different storage treatments, Murwira (1993) found that only anaerobically stored manure from Zimbabwe might improve the availability of N by a reduction of volatilization rates. Probert, Okalebo and Jones (1995) noted that the high pH in Kenyan soils beneath manure piles are conducive to ammonia volatilization. A field study in Zimbabwe, however, found volatilization to be an insignificant source of nitrogen loss (Murwira, Swift & Frost, 1995). The principal mechanisms to reduce ammonia losses during storage appear to be:

- immobilization of ammonium with addition of N-poor materials (Kirschmann & Witter, 1989)
- adsorption of ammoniacal-N with zeolite or clay minerals (Witter & Lopez-Real, 1988)
- lowering pH
- reduction of storage time

- temperature control
- reduction of NH_3 concentration (Dewes, Schmitt, Valentin & Ahrens, 1990)

Nitrogen losses in manure application

Once manure is applied to the field, nitrogen can be lost through numerous pathways. Immobilization of available N is one possible result (Beauchamp, 1986). Nitrogen may be unavailable to plants if the liquid manure is applied to the crop at a non-optimal time (Dewes *et al.*, 1990). One study in Zimbabwe found that early application of manure resulted in more available N for crop growth (Murwira, 1993). Anaerobically-stored liquid manure was found to have high rates of immobilization during application (Kirschmann & Lundvall, 1993). Other studies suggest that early application of manure is necessary to encourage significant N mineralization in time for plant growth. Denitrification may also be an important process of N loss in some cases (Beauchamp, 1986). In well-drained soils with low soil moisture conditions, denitrification may be minimal (Murwira & Kirschmann. 1993). Tropical soils often have lower cation exchange capacities and lower organic matter contents than temperate soils, which means that less adsorption of NH_4^+ can occur. Higher tropical temperatures, evaporation potential and soil water contents during the rainy season may make a conducive environment for ammonia volatilization. The higher overall acidity of soils, however, may lead to lower rates of volatilization.

If manures are not plowed under, NH_3 volatilization can occur (Barry, 1996). Between 20% and 80% of ammoniacal-N can be lost within the first week of manure application (Crane, Westerman & Overcash, 1981). The depth below the soil surface and the presence of NH_4^+ adsorbing constituents such as organic matter or clay minerals both influence volatilization rates (Freney, Simpson & Denmead, 1983). Initially, pH increases with manure application due to high concentrations of NH_4^-N, which increases volatilization rates (Crane, Westerman & Overcash, 1981). Nitrate concentrations appear to be depressed during the initial application of liquid manure, due to saturated conditions (Paul & Beauchamp, 1989). Freney, Simpson and Denmead (1983) noted that the timing of application is also significant for volatilization rates, as plants with more extensive root systems can take up more NH_4^+. Diluted manure slurries may have significantly less volatilization than undiluted slurries (Pain & Thompson, 1989).

Implications for manure extract use

Extensive research has been done on the use of liquid manure in North America. Liquid manure here refers to manure which is diluted with water during cleaning procedures on large farming operations. Although it differs from manure extract, in that there is no separation or steeping process and it is more concentrated, there are many similarities between the substances. Liquid manure contains readily soluble nitrogen fractions, both inorganic (ammoniacal-N) and organic (van Dijk & Sturm, 1983). If stored under anaerobic conditions, inorganic N increases (Kirschmann & Witter, 1992). Manure extract, if it is not stirred, may have increased levels of inorganic N, which would make it effective as a fertilizer. Liquid manure, if it is covered while soaking and has a small surface area, has lower volatilization rates during storage than solid manure. The temperature and lack of aeration in the liquid manure is also important to prevent volatilization. Storage methods of manure extract will be important to ensure value as a N fertilizer. Once manure extract is applied to the soil, the exposure to the atmosphere and high air temperatures prevailing in the dry season is likely to cause ammonia volatilization. In sandy soils, this volatilization may be even higher, because the cation exchange capacity of the soil will be minimal. If the extract is incorporated immediately, this gaseous loss can be minimized. Some of the NH_4^+ may be immobilized in the soil, particularly if the soil has a high C:N ratio (Kirschmann & Witter, 1989). Denitrification and leaching may occur if the soils are watered frequently, and a high moisture level is maintained (Murwira & Kirschmann, 1993). Thus, it is not clear that manure extract will reduce N losses in storage and application or will provide more N to the crop.

The Study

Selection of farmers

The selection of farmers is critical in determining if on-farm research will result in useful information for a given group of people (Werner, 1993; Reinjtnes, Haverkort & Waters-Bayer, 1992). Because the research was intended to help increase food security for resource-poor farmers, we wanted to carry out participatory on-farm experiments with resource-poor farmers. Our focus was women farmers, who have generally been neglected by agricultural research in

Malawi (Ferguson, 1994; Riley, 1995). We decided to draw upon the experience of the Primary Health Care (PHC) programme of Ekwendeni Hospital. We met with a PHC programme nurse, who was able to identify food-insecure families through her contacts at malnutrition clinics, housing, food aid and clothing programmes. From those families, we chose eight women who were interested in carrying out on-farm experiments. This approach could be considered the "key informant" approach in farmer selection (Werner, 1993).

Experimental design

We wanted to incorporate the farmers' perspectives, as well as their current practices and available resources, so that the experiments would be appropriate for their circumstances (Werner, 1993). The experimental design was determined through a combination of discussion with the women, examination of food security interviews conducted initially, and the literature.

We also wanted to ensure that we were using a controlled scientific method, so that the experiment could be repeated and statistically analyzed (Ashby, del Pilar Guerrero, Quiros, Roa & Beltran, 1995). We therefore tried to use a directed participatory approach, where researchers incorporate the farmers' practices and opinions, but control the overall experimental design (Okali, Sumberg & Farrington, 1994).

After being introduced to these women, we interviewed them regarding food security and agricultural methods. We learned that all of the women planned to have a vegetable garden with *Brassica spp.* of either rape (*Brassica napus*) or mpiru (*Brassica juncea*). Their small gardens have about eight beds, each bed 60cm by 100cm (or 0.6 m^2). They have access to some manure (goat, pig, cattle or slaughterhouse) from relatives, neighbours or public locations. They typically receive a half a tin of manure for each vegetable bed, which is equivalent to approximately 9.5 litres. They normally add the manure to the vegetable garden two weeks before or after transplanting. If the manure is added before transplanting, the garden is watered for two weeks to encourage decomposition. Adding the manure two weeks after transplanting allows the plant to establish firmly. Two weeks later, a second addition of manure is done.

Based on this information, we designed a very simple trial to be replicated on each garden. Three treatments were selected:

- Two applications of Solid Manure (SM)

- Two applications of Manure Extract (ME)
- No manure application (Control)

It was assumed that half a tin of manure was available for each vegetable bed. Eight beds were used for each experiment, with two beds for each treatment. All beds would be the same size, 60cm by 100cm. Each bed would hold eight plants, evenly spaced. The transplanted vegetables came from the same nursery. There would be two applications of each treatment: one application at two weeks after transplanting, and one application at four weeks after transplanting. For each application, 1/4 of a tin of manure was available. To make the manure extract, 1/4 of a tin of manure was added to 3/4 tin of water. The beds were randomly assigned a treatment. The experiment was a Randomized Complete Block Design. The experiments were designed to ask three basic questions:

1. How does manure extract compare with manure as a fertilizer for mpiru or rape?
2. Is it worthwhile for food-insecure farmers to make manure extract for their vegetable gardens, rather than using solid manure?
3. What observations and comments do food-insecure farmers have about manure extract?

Site locations

Four of the women did the experiments in their own gardens, and three of the women did a group experiment in one garden, totalling five trials. These trials were replicated at the Development Department experimental farm. The women used a variety of containers for the manure extract: clay pots, metal tins, oil drums or pits lined with plastic. Three of the five trials were not completed because chickens destroyed the plants, but information was collected on their manures and manure extracts. The remaining two experimental sites were located within five kilometres of Ekwendeni, at an altitude of approximately 1,200 metres above sea level (Young & Brown, 1962). The mean annual rainfall is approximately 840 to 1,015mm, and the mean annual temperature ranges between 18.3 and 21 C (Kingston, Mitchell & Ntokotha, 1978). There was no rainfall recorded at a nearby climate station during the experimental period. The soils are a sandy loam and sandy clay loam, with a pH range of 7.85 to 8.06. They correspond most closely to the Class I of the Ekwendeni series (Kingston, Mitchell & Ntokotha, 1978).

Sampling and analytical procedures

Soil was sampled twice during the trials. The untreated soil was sampled at each experimental location. A soil sample to a depth of 30cm was taken with a spade in five randomly selected locations of each garden. The samples were placed in a tin, mixed thoroughly, and approximately 1 kg of soil was taken from this mixture. The soil was air-dried for 24 hours, sealed in a polyethelene bag and transported to the laboratory. After the plants were harvested, the soil was sampled again, but this time a sample was taken from each bed, rather than for the entire garden.

The manure was sampled twice during the experiment. Prior to each application, a fresh manure sample was taken for each experimental site. The manure was stirred thoroughly and a sample of approximately 1kg was taken, sealed in a polyethelene bag and frozen. The frozen samples were taken to the laboratory. The manure extract was sampled several times during the experiment. Ten minutes after the manure sack had been placed in the container of water, a 10ml sample was taken with a clean syringe, and put in a sealed container. The container was frozen. A sample was taken approximately every two days. Each time a sample was taken, pH and temperature were also measured. Approximately seven manure extract samples were taken prior to application. The sampling was repeated for the second manure extract application. At application, a homogeneous sample was taken from the manure sack, sealed in a polyethelene bag and frozen.

Plant samples were collected at harvesting. The women farmers used knives to harvest leaves from the plants, keeping the leaves from each treatment bed separate. A random selection was taken from each bed, ensuring that the leaves were not contaminated by dust, damaged by insects, mechanically injured or diseased. The leaves were washed with water and dried in an oven at approximately 65C within 24-48 hours of collection. These collection and preparation procedures are in accordance with Soil Science Society of America standards (Jones & Case, 1990).

All the plants were measured during the growth period – at transplanting, approximately two weeks after transplanting and at harvesting. A ruler was placed at the base of the plant and the height of the tallest leaf was recorded. The presence of disease, pests or plant withering was also recorded at this time.

The soil samples were analyzed for total N, NH_4^+, NO_3^- , pH, organic matter content and particle size. The manure samples were analyzed for total N, NH_4^+, NO_3^- and total P. The manure extract samples were analyzed for NH_4^+

and NO_3^-. Plant samples were analyzed for total N. Standard analytical techniques were used for this laboratory work – modified Kjeldahl N-analysis, P colorimetric tests, K, Ca, Mg, Zn and Cu using atomic adsorption techniques (McGill & Figueiredo, 1993). NH_4^+, NO_3^- analyses were done using the Technicon Auto Analyzer (Keeney & Nelson, 1982). Particle size analysis was done using the pipette method (Kilmer & Alexander, 1949).

Farmer evaluation

In addition to the items named above, we also wanted to ascertain what factors the farmers felt were important to measure. Several meetings were held with the participating farmers to design an evaluation process for the trials. The factors which they listed as important include: size, colour and number of leaves, presence of pests or disease, and growth. The evaluation attempted to incorporate these factors in quantitative and qualitative ways. The farmers made the evaluations at harvesting time. A farmer initially ranked all the beds according to all the criteria that she had listed. She then harvested each bed separately, placing all the leaves on a piece of paper labelled with the bed number. The woman then sat away from the garden, and we brought one harvested bed of leaves to her at a time. This method made the second part of the evaluation "blind", because the farmer did not know which bed she was evaluating. A matrix score system was used to incorporate the criteria listed by the farmers (Werner, 1993). The leaves were divided into three groups for colour: very green, (3 points), green (2 points) and yellow (1 point). A "colour score" was determined by multiplying the number of leaves by their score, and adding up the total. The number of size groups was determined by the farmers, and the average length (cm) of the leaves was recorded. By multiplying the number of leaves by their length, and adding up the total, a size score for each bed was determined. Farmers also noted the number of leaves with aphids or disease present, and weighed all the leaves from one bed together to get a total weight.

RESULTS

Solid manures

The manures collected generally had low to medium nitrogen contents, a fairly low phosphorus content, and typical contents of other nutrients in comparison to other studies in the region (Murwira, Swift & Frost, 1995; Probert, Okalebo & Jones,

1995). Zinc and copper were generally at very low levels. Raw data, arithmetic means and standard deviation of N, P, K, Ca, Mg, Zn and Cu are given in Table 8.1.

8.1 Chemical composition of 12 manures collected in Ekwendeni district, June 1997. (See opposite)

Mineral (or inorganic) nitrogen (N) was measured at the time of manure collection and again at the time of application. Two of the farmers (sites B and C) collected the manure almost two months prior to application. They stored the manure in bags indoors. A significant increase in mineral N (primarily nitrate NO_3^-) was measured for the manure stored in bags during this storage period. The mineral N was almost entirely ammonium at collection, but at application the majority of mineral N was nitrate. Mineralization occurred in the manure kept in bags without substantial losses through leaching, denitrification or volatilization. At the Development Department's Demonstration Farm, some of the goat and slaughterhouse manure was stored in heaps outside. This manure had lower levels of mineral N at application, most likely due to volatilization, as there had been no rainfall which would have stimulated denitrification.

Manure extracts

- **pH:** The pH of the manure extracts rose initially from the close to neutral pH of the water, and then stabilized over the two week period. Typical pH values were between 7.5 - 8.5. The pH levels varied from site to site, which may be due to manure composition, water pH and the type of container used. At two sites, the pH rose to over 9, but the majority of pH values were closer to 8. The pH remained high enough, however, to mean that some of the nitrogen would be lost through ammonia volatilization. In order to avoid nitrogen losses as ammonia, the pH must remain neutral or lower.
- **Mineral nitrogen:** During the two weeks the manure was steeping, mineral nitrogen increased in the liquid portion of the manure extracts. At sites A and C, the total amount of mineral N increased and then decreased, but did not fall below the starting level. The total amount of mineral N decreased for the liquid extract at site F. Samples were lost for sites B, D and E. The total amount of mineral N in the liquid extract varied from a low of about 10 mgN/l to a high of over 100 mgN/l. One of the major sources of difference was in the containers used. In the sites where a pit was dug, there were lower

8.1 Chemical composition of 12 manures collected in Ekwendeni district, June 1997.

Sample Description	Site	Collection Location	%	μg/g or mg/kg						
			Total N	Total P	Total K	Total Ca	Total Mg	Total S	Cu	Zn
Fresh goat	A	Covered *khola*	2.07	4000	8300	17600	1600	1600	8	7
Fresh goat	B	Covered *khola*	1.91	4900	7800	15800	1500	2200	19	8
Older cattle	B	Outdoor *khola*	0.90	3300	10000	18500	2500	1800	14	9
Fresh goat	C	Covered *khola*	0.92	2000	14800	19100	1900	1400	11	7
Older pig	C	Outdoor *khola*	0.34	2200	16300	9900	300	300	18	8
Fresh slaughterhouse	C	Slaughterhouse (outdoor)	0.78	4000	6800	8300	300	800	9	11
Fresh slaughterhouse	D	Slaughterhouse	0.70	3600	13500	15900	300	1700	6	64
Fresh slaughterhouse	E	Slaughterhouse	0.49	1800	13300	10100	300	600	7	7
Fresh goat	F	Outdoor *khola*	2.13	3800	10800	14000	800	1900	21	5
Older cattle	F	Outdoor *khola*	0.50	1400	10300	10100	300	900	21	5
Fresh slaughterhouse	F	Slaughterhouse	0.44	2500	5000	18300	400	900	7	15
Older pig	F	Ditch	0.55	3900	9000	15800	800	1200	13	8
Mean			0.97	3100	10500	14500	900	1300	13	13
S.D.			0.61	1000	3100	3600	700	500	5	15

levels of mineral N in the manure extract. There appeared to be more volatilization due to the increase in surface area. The total amount of mineral N lowered at some sites. The drop may be attributed to a combination of an increased concentration of ammonium, high pH, and high temperatures, all of which influence ammonia gas losses.

At the time of application to the plots, the manure extract had significantly less mineral nitrogen than the solid manure. However, after the two weeks of soaking, the solid portion remaining in the bag ("bag manure") contained a significant amount of mineral N. Manure extract is almost entirely ammonium, while the solid manure is almost entirely nitrate.

Yields

Nitrogen uptake appears to be one of the causes in yield differences. Plant uptake of nitrogen was significantly higher for solid manure treatments, which had significantly higher yields than both the manure extract and the control for the two sites that completed the experiments (Table 8.2). Soil samples taken at the second harvest also indicated a slight correlation between soil mineral nitrogen and yield.

Table 8.2 Mean yield, plant N uptake and soil mineral N at harvest

Treatment	Mean Yield (g/m^2)	Plant N Uptake (N g/m^2)	Soil Mineral N (mg/kg)
Solid (SM)	1956*	86*	30
Extract (ME)	1005*	43	26
Control (C)	577	25	14

* Indicates a significant difference to the control at a 95% probability level.

Discussion

The manures available to the farmers were not excellent nutrient sources, but provided some nitrogen and other nutrients for their crops. Managing manure to conserve nitrogen may be a way to increase crop production. Manure extract appears to stimulate the production of mineral N, similar to liquid manure anaerobic processes in storage. High pH levels, and in some cases the containers' high surface areas, however, induce ammonia volatilization. The liquid in the manure extract did not have amounts of mineral N equal to those in the solid manure. The solid portion remaining in the bag also contained substantial amounts of mineral N. Most organizations promoting the use of manure extract do not discuss what to do with the "bag manure". The evidence from these experiments suggests that it should also be used in the garden.

Plant samples taken at harvest showed that solid manure was associated with higher yields than manure extract. This difference is at least partially due to nitrogen differences. Ammonia volatilization may be occurring at application of the manure extract, leading to the difference in yield. Volatilization has been a suggested loss of N for other species such as cabbage (Smith & Hadley, 1988). The manure extract is initially applied to the plant when the root systems are not very extensive, and so cannot take up much NH_4^+ (Freney, Simpson & Denmead, 1983). Denitrification, due to the high levels of moisture at application, may also be a source of N loss for manure extract, but because this moisture level is very temporary, it may not be the main source of N loss. Another possibility is immobilization of ammonium in the soil. Most of the manures used, however, contained few sources of decomposable C, such as litter or bedding, although the slaughterhouse and pig manure did contain some soil and grass. Without a source of readily decomposable carbon, immobilization is not likely to occur. It may be that the manure extract was applied at a time when the plants did not take up much N, although these *Brassica* species, like cabbage, are relatively efficient at utilizing available N (Zebarth, Freyman & Kowalenko, 1991). An earlier application of manure extract may encourage N mineralization in time for plant growth (Murwira, 1993).

In their evaluations of the manure extract experiments, the farmers noted that the technique saved them substantial labour, because they did not have to water the garden before transplanting the vegetables. Several farmers felt that manure extract should be tested as an alternative fertilizer in their maize gardens. Although they noted the yield differences between solid manure and manure extract, they felt that manure extract produced satisfactory yields in their vegetable gardens.

There are a number of manure management practices the farmers might try to improve vegetable yield and quality. They should try to collect the manure just prior to use, as fresh manure contains the highest quantities of nitrogen. The farmers' time and labour constraints, however, may mean they have to store the manure for some time. In order to prevent nitrogen leaching and ammonia losses, the farmers should store the manure in sealed bags indoors. The practice of adding the manure several weeks before transplanting vegetables should be encouraged, because it increases mineralization. The farmers should try to add the manure at least one month before transplanting. If using solid manure, the farmers should try to incorporate the manure into the soil as much as possible. Current practices do not always involve full incorporation of the manure. If applying the manure prior to transplanting, the farmer can use her hoe to fully incorporate the manure. If applying the manure after transplanting, soil or mulch can be used to cover the manure, so that the transplants are not disturbed.

Those farmers who decide to use manure extract in their gardens can try several management practices which may increase its effectiveness as a fertilizer. In order to keep the temperatures of the extract low and to reduce volatilization, the containers should be covered well with grasses and be stored in a shady area. Containers with a low surface area, such as tins, clay pots, or deep pits (rather than wide ones), should be used if at all possible. Materials that lower pH can be added, such as tea residue, maize bran[2] or beer residue. Application of the manure extract might be tried two or three times:

- a month before transplanting, to encourage mineralization
- two weeks after transplanting, when the plant has a high N requirement
- three to four weeks later, when the root system is quite extensive

Finally, the manure extract should be incorporated rapidly and fully into the soil. The farmer might put a mulch (e.g. leaves, maize stalks, rocks, soil) over the area after applying liquid manure, to prevent ammonia losses. Alternatively, the farmer might dig a small trench or hole beside the plant and pour the manure extract into it. The trench or hole can then be filled with soil.

Endnotes

1. Malawians refer to this fertilizer type as liquid manure. Liquid manure in North America, however, refers to liquified manure from large farm operations. Some organic agriculture organizations call this method 'manure tea'. I have called the amendment 'manure extract'.
2. Maize bran is the residue from pounding maize. It has a fairly low pH.

References

Ashby, J.A.T., del Pilar Guerrero, G.M., Quiros, C.A., Roa, J.I., and Beltran, J.A. 1995: *Institutionalizing farmer participation in adaptive technology testing with the*

"CIAL". Colombia: CIAT.

Barry, D.A.J. 1996: Nitrogen and Carbon Transformation in Conventionally-Handled Livestock Manures. In M. Goss, J.D.P. Stonehouse and J.C. Giraldez (eds.), *Managing Manure for Dairy and Swine*, Ch. 4. Guelph: University of Guelph.

Beauchamp, E.G. 1978: Ammonia volatilization from sewage sludge applied in the field. *J. Environ. Qual*, 7 (1), 141-146.

Beauchamp, E.G. 1986: Availability of nitrogen from three manures to corn in the field. *Can. J. Soil Sci*, 66, 713-720.

Beauchamp, E.G., Kidd, G.E. and Thurtell, G. 1982: Ammonia volatilization from liquid dairy cattle manure in the field. *Can. J. Soil Sci*, 62, 11-19.

Beauchamp, E.G. and Paul, J.W. 1989: A Simple Model to Predict Manure N Availability to Crops in the Field. In J.A.A. Hansen and K. Henriksen (eds.), *Nitrogen in Organic Wastes Applied to Soils*, pp. 140-149. London: Academic Press.

Carr, M. (ed.). 1991: *Women and Food Security: the experience of the SADCC countries.* London: IT Publications.

Crane, S.R., Westerman, P.W. and Overcash, M.R. 1981: Short-term chemical transformations following land application of poultry manure. *Transactions of the ASAE*, 24, 382-390.

Dewes, T., Schmitt, L., Valentin, U. and Ahrens, E. 1990: Nitrogen losses during the storage of liquid livestock manures. *Biol. Wastes*, 31, 241-250.

Ferguson, A. 1994: Gendered Science: A Critique of Agricultural Development. *American Anthropologist*, 96, 540-552.

Freney, J.R., Simpson, J.R. and Denmead, O.T. 1983: Volatilization of ammonia. In J. R. Freney and J.R. Simpson (eds.), *Gaseous Losses of Nitrogen from Plant-Soil Systems*, pp. 1-32. The Hague: Martinus Nijhoff.

Hiwa, S.S. 1988: *Agriculture Development Policy and Food Production*. Zomba: Centre for Social Research, University of Malawi.

Jones, J. and Case, V.W. 1990: Sampling, Handling and Analyzing Plant Tissue Samples. *Soil Testing and Plant Analysis*. Madison: SSSA.

Kasembe, J.N.R., Semoka, J.M.R. and Samki, J.K. 1983: Organic farming in Tanzania. In J.M.R. Semoka, F.M. Sha, R.R. Harwood and W.C. Liebhart (eds.), *Resource-efficient farming methods for Tanzania*, pp. 36-40. Emmaus: Rodale Press.

Keeney, D.R. and Nelson, D.W. 1982: Nitrogen: inorganic forms. In A. Page, R. Miller and D. Keeney (eds.), *Methods of Soil Analysis Part II*, 2nd ed, pp. 643-698. ASA, SSSA.

Kilmer, V.J. and Alexander, L.T. 1949: Methods of making mechanical analyses of soils. *Soil Science*, 68, 15-25.

Kingston, J.D., Mitchell, A.J.B. and Ntokotha, E.M. 1978: *Soil Surveys* 1971-72. Zomba: Ministry of Agriculture and Natural Resources, Soil Survey Section, Government of Malawi.

Kirchmann, H. and Witter, E. 1989: Ammonia volatilization during aerobic and anaerobic manure decomposition. *Plant and Soil* 115:35-41

Kirschmann, H. and Witter, E. 1992: Composition of fresh, aerobic and anaerobic farm animal dungs. *Bioresource Technology*, 40, 137-142.

Kirschmann, H. & Lundvall, A. 1993: Relationship between N immobilization and volatile fatty acids in soil after application of pig and cattle slurry. *Biol. Fertil. Soils*, 15, 161-164.

MacLean, A.J., Miller, M.H. and Robinson, J.B. 1983: The fertilizer potential of animal manures and environmental constraints on their use. *Farm Animal Manures in the Canadian Environment*. Ottawa: NRCC.

McGill, W.B. and Figueiredo, C.T. 1993: Total Nitrogen. In M.R. Carter (ed.), *Soil Sampling and Methods of Analysis*, pp. 201-211. Lewis Publishers.

Msukwa, L.A.H. 1986: *Agricultural development and undernutrition: the case of Malawi*. Zomba: Centre for Social Research.

Msukwa, L.A.H. 1994: *Food policy and production: towards increased household food security*. Zomba: Centre for Social Research.

Mughogho, S.K. 1989: *Soil fertility research in Malawi: a review*. Lilongwe: Bunda College of Agriculture.

Mughogho, S.K. and Ngwira, M.E. 1996: *Soil fertility bibliography for maize-based cropping systems in Malawi*. Lilongwe: Soil Fertility Network, Bunda College of Agriculture.

Murwira, H. 1993: Synchrony relationships of nitrogen release and plant uptake in a Zimbabwean sandy soil amended with manure and fertilizer nitrogen. *African Crop Science Journal*, 2 (1), 69-77.

Murwira, H.K. 1995: Ammonia losses from Zimbabwean cattle manure before and after incorporation into soil. *Tropical Agriculture*, 72(4), 269-273.

Murwira, H. and Kirschmann, H. 1993: Carbon and nitrogen mineralization of cattle manures, subjected to different treatments, in Zimbabwean and Swedish soils. In K. Mulongoy and R. Merckx (eds.), *Soil Organic Matter Dynamics and Sustainability of Tropical Agriculture*, pp. 189-198. London: John Wiley & Sons.

Murwira, H., Swift, M.J and Frost, P.G.H. 1995: Manure as a key resource in sustainable agriculture. In J.M. Powell, S. Fernandez-Rivera, T.O. Williams and C. Renard (eds.), *Livestock and Sustainable Nutrient Cycling in Mixed Farming Systems of sub-Saharan Africa*, pp. 131-148. Volume II: Technical Papers. Addis Ababa: ILCA.

Okali, C., Sumberg, J. and Farrington, J. 1994: *Farmer participatory research: rhetoric and reality. London: IT Publications.*

Pain, B.F. and Thompson, R.B. 1989: Ammonia volatilization from livestock slurries applied to land. In J.A.A. Hansen and K. Henriksen (eds.), *Nitrogen in Organic Wastes Applied to Soils*, pp. 202-211. London: Academic Press.

Paul, J.W. and Beauchamp, E.G. 1989: Relationship between Volatile Fatty Acids, Total Ammonia and pH in Manure Slurries. *Biological Wastes*, 29, 313-318.

Price, M. 1985: How Adequate is Chicken Manure Tea as a Fertilizer? *Echo Technical Note G- 3*. (From a Research Report by N. Duddles, California Polytechnic University).

Probert, M.E., Okalebo, J.R. and Jones, R.K. 1995: The use of manure on smallholders' farms in semi-arid Eastern Kenya. *Expl. Agric*, 31, 371-381.

Reinjtnes, C., Haverkort, B. and Waters-Bayer, A. 1992: *Farming for the Future: an introduction to Low-External-Input and Sustainable Agriculture.* Netherlands: Macmillan Press.

Riley, P.J. 1995: Gender Issues and the Training of Agricultural Extensionists in Malawi. *Agriculture and Human Values*, 12, 31-38.

Saka, A.R., Bunderson,W.T., Itimu, O.A., Phombeya, H.S.K and Mbekeani, Y. 1994: The effects of *Acacia albida* on soils and maize grain yields under smallholder farm conditions in Malawi. *Forest Ecology and Management*, 64, 217-230.

Smith, S.R. and Hadley, P. 1988: A comparison of the effects of organic and inorganic nitrogen fertilizers on the growth response of summer cabbage (*Brassica oleracea* var. *Capitata cv. Hispi F1.). J. Horticulture Science*, 63(4), 615-620.

Spring, A. 1988: Using Male Research and Extension Personnel to Target Women Farmers. In S.V. Poats, M. Schmink and A. Spring (eds.), *Gender Issues in Farming Systems Research and Extension*, pp. 407-426. Boulder: Westview Press.

UNICEF and Government of Malawi. 1987: *The Situation of Children and Women in Malawi*. Lilongwe: UNICEF.

United Nations Development Program (UNDP). 1997: *Human Development Report 1997*. Oxford: Oxford University Press.

Van Dijk, T.A. and Sturm, H. 1983: Fertilizer Value of Animal Manures on the Continent. Paper read to the Fertilizer Society of London, 1st Dec. 1983, Greenhill House, London.

Vlek, P.L.G. 1993: Strategies for sustaining agriculture in sub-Saharan Africa: the fertilizer technology issue. In J. Ragland and R. Lal (eds.), *Technologies for Sustainable Agriculture in the Tropics*, pp. 265-278. Madison: ASA, CSSA and SSSA.

Werner, J. 1993: *Participatory Development of Agricultural Innovations.* Eschborn: GTZ.

Witter, E. and Lopez-Real, J. 1988: Nitrogen losses during the composting of sewage sludge, and the effectiveness of clay soil, zeolite and compost in adsorbing the volatilized ammonia. *Biological Wastes*, 23, 279-294.

Young, A. and Brown, P. 1962: *The physical environment of Northern Nyasaland.* Zomba: Government of Nyasaland.

Zebarth, B., Freyman, J.S. and Kowalenko, C.G. 1991: Influence of nitrogen fertilization on cabbage yield, head nitrogen content and extractable soil inorganic nitrogen at harvest. *Canadian Journal of Plant Science*, 71, 1275-1280.

CHAPTER 9

AGROFORESTRY WITH RAINWATER HARVESTING IN BOTSWANA

by Margriet Berkhout

BACKGROUND

Botswana has a semi-arid climate. Average annual rainfall varies from less than 250 mm in the southwest of the country to over 650 mm in the north. Rainfall is erratic and mainly concentrated in a rainy season lasting from October until April. Droughts are common and often last for several years.

The population is about 1.3 million, with an annual growth of 3.5 percent. About 80% of the population lives in the eastern part of the country in what is called the hard veld. The soils there are ferruginous: well drained loam soils with moderate fertility. The rest of the country consists of sand veld, with very deep sandy soils of low fertility.

Most Batswana live in rural areas. Agriculture is primarily subsistence in nature and can be described as extensive or semi-intensive mixed farming. Most farmers own less than 10 cattle which serve as a source of status, drought insurance, pension and income. Crop production is the prime source of food for most rural Batswana, especially for the 26% of households that do not own cattle.

The major crops are sorghum, maize, pulses and watermelons. Most years, the majority of households do not produce enough food for subsistence needs. Constraints on crop production include low inputs and management, along with unfavourable agroclimatic and soil conditions. There are several government assistance and incentive programmes that assist farmers with inputs such as draft power, fencing and relief to sustain households through drought periods. However, these subsidies have created a "handout mentality" and have not favoured self-help and self-reliance. With financial assistance for ploughing and planting, but not weeding, large areas are cultivated and then left without management after the money is received. Currently, after two very good rainy seasons, these programmes have been scaled down.

Agroforestry by Veld Products Research

Veld Products Research (VPR) has been investigating and promoting the potential of veld products, or non-timber forest products, for increasing food security and income generation since 1989. Much emphasis is on the research and domestication of indigenous fruit trees. The VPR agroforestry project took a further step, by integrating indigenous fruit trees into crop production. Fruit trees are often left when fields are cleared for crop production, so the basic agroforestry concept is not new. It is the concept of actively growing trees and managing them as an integrated part of the farming system that is still novel in most of Botswana.

In 1995, a pilot trial was established at the VPR site in Gabane. This trial combined traditional crops with indigenous fruit trees and rain water harvesting. The choice of these components was justified as follows:

- Traditional crops are adapted to the Botswana environment and are familiar and acceptable to subsistence farmers.
- Rainwater harvesting maximizes the availability of rainfall to crops which improves yields and could prevent total crop failure in poor rainfall years. At the same time, it controls soil erosion by intercepting run-off.
- Most indigenous fruit trees produce fruit even in drought years. Indigenous fruit is popular with rural and urban people alike, but the regeneration of the fruit trees is failing, due to overgrazing. As a result, many indigenous fruit trees are disappearing from the village surroundings, or have disappeared already. Indigenous fruit production could improve farm household nutrition and food security, as well as generate income from the sale of surplus fruit or fruit products.

Establishment and Management of the Trial

The base of the trial consisted of a five-year-old morula (*Sclerocarya birrea*) orchard. The trees were spaced 12 meters by 12 meters. Between the trees, contour ridges were established (see Figure 9.1). Sorghum (*Sorghum bicolor*) and cowpeas (*Vigna unquiculata*) were intercropped along and on top of the ridges. Perpendicular tie ridges, one meter in length, were constructed every two metres along the ridge and served to stem the lateral flow of water. Mogorogorwane (*Strychnos cocculoides*) were planted between the morula trees prior to planting the crops. The distances between the ridges were either six metres (one ridge between tree rows) or four metres (two ridges between tree rows). The area between the ridges served as a catchment to create run-off towards the crops. It is expected that the crops grown in the six-metre catchments will receive more rainfall and have higher yields than the crops in the four-metre catchments. In some of the catchment areas, watermelons (*Citrullus Lanatus*) were grown at a spacing of three metres. By growing the watermelons, the productivity of the system should be increased, but the trade-off may be that they decrease the run-off to the main crops.

Figure 9.1 Layout of the agroforestry trial at Veld Products Reasearch, Gabane, Botswana

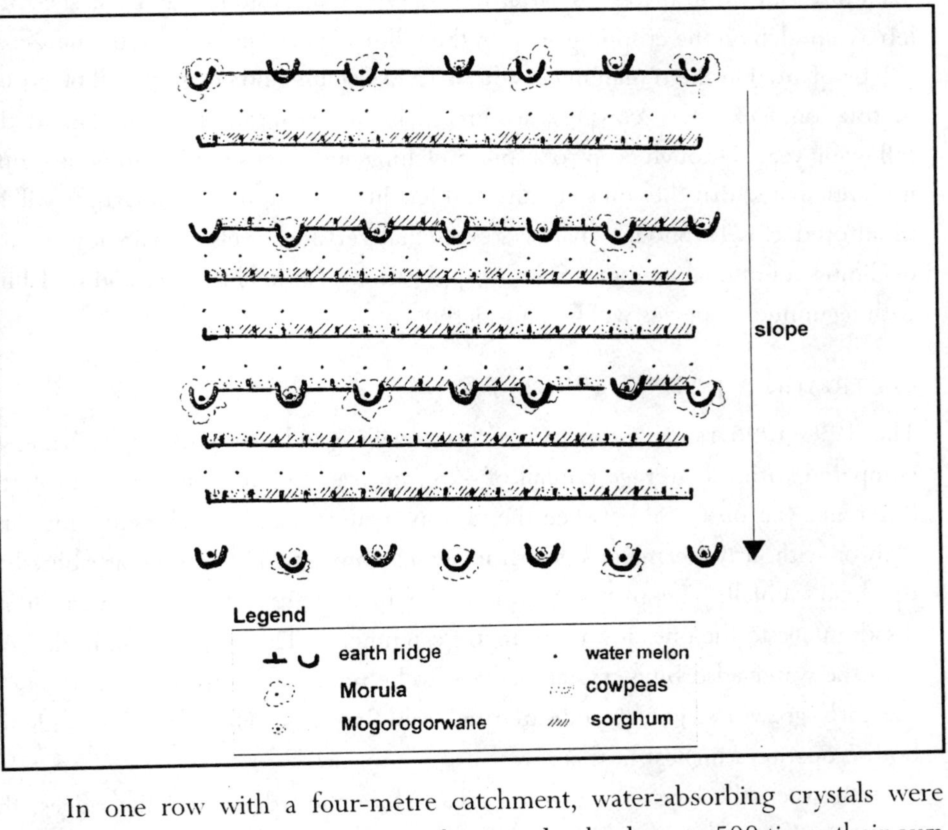

In one row with a four-metre catchment, water-absorbing crystals were incorporated into the cropping area. The crystals adsorb up to 500 times their own volume in water, which is then slowly released to the plants. It is claimed that they also facilitate the plants' nutrient uptake. The crystals were included in the trial to assess their potential for crop production in Botswana. They are quite expensive, and a cost-benefit analysis should reveal if their application is practical.

This configuration has several potential advantages. After constructing the ridges in the first year, no tillage is required, although in the following years the ridges will need to be repaired before planting the new crop. In addition, planting can be done as soon as the rains have started so that the crops can fully benefit from the rains. In the traditional system, ploughing can start only once the soils have softened after sufficient rainfall, when much of the precious rainfall has been lost to the crops. Planting is done in small individual holes. Behind the ridges, the seeds can also be dug in with a hoe.

Regular weeding is necessary to keep the catchment areas free of weeds and maximize run-off to the crops. After the harvest, crop residues can be slashed and left as a mulch on the cropping area. In the following planting season, the new crop will be planted in the remaining residues. The sorghum and cowpeas will be grown in rotation; i.e. where cowpeas are grown in one year, sorghum is planted the following year. Through crop rotation, mulching and the use of leguminous crops, it is anticipated that the soil structure and fertility will improve. Soil fertility will be monitored closely. Should analysis reveal that fertility levels remain low or are declining, additional measures such as application of animal manure and mulching with leguminous species will be considered.

Observations in 1995/1996

The 1995/1996 rainy season was exceptionally good; 709 mm in Gabane, as compared with the average rainfall of 376 mm over the previous ten years. Little difference was observed between the various treatments, i.e. catchment width and with or without watermelons grown in the catchments. This can be attributed to the high rainfall. The plants grown in the four-metre catchments were at no disadvantage to the ones in the six-metre catchments. The crops grown in the row with the water-adsorbing crystals seemed to be under stress from waterlogging in the early growth stages, as indicated by early flowering. However, the yields for both crops were highest in this row.

Between the bays (the planted area before the ridges) and the ridges, the following differences were observed:

- Germination was better in the bays than on the ridges for both crops.
- Flowering was earlier on the ridges for both crops, indicating stress.
- Yields on the ridges were higher than in the bays for both crops.

The differences in germination and flowering can be explained by the fact that the ridges are more prone to desiccation than the bays. However, the plants on the ridges generated higher yields. As the ridges were constructed with topsoil from the bay areas, it is likely that soil fertility and structure on the ridges were better then in the bays. The yields, expressed as kg/hectare, are displayed in Tables 9.1 and 9.2. All treatments included watermelons.

Table 9.1 Yield of cowpeas (kg/ha)

	6m width	4m width	4m width (with water-absorbing crystals)	Control
Bay	147	284	417	
Ridge	353	408	418	
Total	**521**	**698**	**842**	**330**

Table 9.2 Yield of sorghum (kg/ha)

	6m width	4m width	4m width (with water-absorbing crystals)	Control
Bay	92	63	316	
Ridge	311	349	269	
Total	**410**	**415**	**590**	**579**

Many watermelons were lost because of the heavy rains. As a result, the watermelon data are not analyzed. The treatments without watermelons could not be analyzed due to non-orthogonality in the trial design. However, no difference was observed between the crops grown in the catchments with watermelons and without. In a year with poor rainfall, it is expected that the differences between the treatments will be more pronounced.

Because of the good rainfall, the yield of the traditionally ploughed control was good. As the agroforestry trial was designed for poor rainfall years and only 17 to 25% of the area was planted, it was expected that the yields would be lower than the control. This was true of the sorghum, apart from the row with the water-absorbing crystals, but the cowpea yields were much better. It is anticipated that in years with less rainfall, the advantage of this agroforestry system with rain water harvesting will be greater for all crops, including the sorghum.

The crops growing directly under the morula trees had lower yields as compared to those growing further away. The 11 morula trees produced 34 kg of fruit. The trees have all been grafted and started producing in 1994. The graft failed on one tree; it reverted to rootstock and does not produce any fruit. The grafted trees start bearing fruit after three to four years, but early production is low. Female ungrafted trees only start production after 10 to 12 years.

At the writing of this paper in 1997, the second year's harvest had not been completed, therefore only indicative observations can be given. This year, there seems to be little difference between the crops grown under the trees and those grown further away. The lower branches of the trees were pruned last winter, which reduced shading of the crops. The appearance of the crops suggests that the main constraint to their productivity was shading under the trees last year, and not, or less so, competition for nutrients.

The continued high rainfall, with almost 1,000 mm recorded in Gabane so far, has resulted in the loss of many watermelons and cowpeas. The humid conditions also favoured a high incidence of aphids in the cowpeas in both years. It was interesting to note that the aphids did not spread, because of the intercropping pattern and the catchment areas.

Pest and Nutrient Problems

Grain-eating birds posed a major problem to the sorghum yield. In 1996, several methods of discouraging the birds were tried: of these, a fake hawk was moderately successful. This year, however, the effectiveness of this method lasted only a few days. The only successful method seems to be the continuous presence of people to scare the birds away when the sorghum is in its most vulnerable stage. That was not possible this year, and most of the sorghum yield was lost. The trial at the VPR site was particularly vulnerable to bird damage. This is because it was planted early and produced the first grain in the area and because the trial is located on a forested hill which is a bird habitat. VPR recognizes that trees in a field offer shelter to birds and may attract them, which could be a main constraint in the adoption of agroforestry systems. However, on this site, the results would have been the same even if there were no trees within the trial area. In the next season, VPR intends to use sorghum varieties which are resistant to indigenous birds.

Damage to the crops by insects such as aphids and blister beetles was not a major problem in the last two years. In the next season, the VPR will experiment with natural insecticides/fungicides such as the neem (*Azadirachta Indica*) and indigenous plants with similar properties.

The bays are deficient in nutrients, notably nitrogen, as the topsoil is used to construct the ridges. Mulching and rotation with a leguminous crop (cowpeas) is expected to improve the soil, but may not be sufficient to restore and improve soil fertility. Should soil analysis indicate that this is the case, additional measures will be taken, such as the application of manure, or additional mulching material from leguminous species. This could be obtained from a leguminous hedge that will be grown around the plot.

Future Directions - the Community Approach

The current trial was established to see if this agroforestry system would have potential to improve food and household security. Now that the indications are that it has, VPR wants to further develop the system in participation with communities. The Community Based Agroforestry Project (CBAFP) will start this year, with demonstration plots at schools in five different villages in the northeast and east of Botswana. After the first harvest of these plots, a Participatory Rural Appraisal (PRA) will be held in all the villages to establish how agroforestry could best be integrated in the local farming systems and to identify farmers who are interested in participating in on-farm plots. The demonstration plots will be based on the trial in Gabane, but with two more species of indigenous fruit trees, more crops and without the water-absorbing crystals. The choice of species and the design and management of the on-farm plots will be subject to the preference of the farmers. They will be encouraged to make modifications according to their experience and ideas. Through continuous consultation with the farmers, the schools and agricultural extension officers, and through activities in the community, VPR hopes to develop its agroforestry system into systems that are sustainable and feasible, and that will be adopted by subsistence farmers.

The CBAFP will run for two years and is considered the first phase of a long-term project. The follow-up will depend on the farmers' response during the first two years. Later, the project may expand in other parts of the country. In the near future, VPR intends to investigate the processing of indigenous fruit at village level and to establish a buying and marketing network. The agroforestry project will be linked with these activities.

CHAPTER 10

MYCORRHIZAL CONTRIBUTION TO THE ESTABLISHMENT OF FRUIT TREES IN SOUTHERN AFRICA

by Stanley Mateke, Terence P. McGonigle and Robert C. Sinclair

INTRODUCTION

Southern Africa is faced with ever-increasing demands for food production to keep pace with a fast-growing population (McCullum, 1994). However, more than 50% of Southern Africa is unsuitable for cultivation because of insufficient rainfall. In cultivated regions, a lack of soil fertility is a common problem. In addition, high population densities in localized areas have caused declining productivity and soil degradation in many grazing lands (Booth, 1994). A number of different approaches have been taken to try to deal with these challenges. The domestication of indigenous fruit trees is one approach with considerable potential as an alternative form of agriculture in semi-arid areas. Many indigenous fruit trees will produce on marginal soil and in drought years. The propagation of indigenous fruit trees is being developed at Veld Products Research (VPR), a non-governmental organization in Gaborone, Botswana. Studies at VPR have shown that when seedlings of various indigenous fruit trees are transplanted into field plots, they often grow poorly and they frequently die within a few months. The purpose of the three companion studies described here was to contribute to the tree-propagation effort by studying the biological interactions between these trees and the mycorrhizal fungi found in the soils where the trees grow in the veld.

Before beginning out-planting trials, more information was needed about the occurrence of mycorrhizae on these trees in their natural condition. The assumption was made that the mycorrhizal associations these fruit trees make when they propagate successfully in the wild would be well suited to promote success in new plantings. The aim of Study A was to document the occurrence of mycorrhizal morphotypes on tree specimens as they occur naturally in the veld. Study B was conducted to describe the mycorrhizal characteristics of plants found in association with the fruit trees. The natural inoculum occurring on these plants might be used to supply appropriate mycorrhizae to seedlings that are transplanted alongside them. Study C comprised fruit tree out-planting trials.

Mycorrhizae

Mycorrhizae are naturally occurring associations between roots and certain soil fungi (Allen, 1991). The combination of the root and the fungus together constitutes the mycorrhizae, which literally means "fungus-root".

It has been known for some time that in both temperate and tropical systems, the establishment of tree seedlings is often of very limited success unless significant quantities of soil from beneath appropriate vegetation is introduced along with the seedlings. Mycorrhizae are associations which benefit both participants (Smith & Read, 1996). Two main types of mycorrhizae are distinguished. Ectomycorrhizae form on conifers, oaks, and some tropical legume trees. Arbuscular mycorrhizae form on grasses, forbs and the majority of tropical trees. The fungi formed on the roots of the trees of interest in this study were expected to be of the arbuscular mycorrhizal type. Arbuscular mycorrhizae (AM) contribute to plant growth by helping with phosphorus uptake from the soil into the plant (Hayman, 1983), and by reducing the occurrence of root diseases (Marx, 1969; Newsham, Fitter & Watkinson, 1995). In return, the fungus takes sugars from the tree. These sugars are usually in excess of the tree's needs. The dominance of the arbuscular mycorrhizal association in plants in arid and semi-arid habitats is well documented (McGee, 1986; Mejstrik & Cudlin, 1983; Miller, 1979; Mukerji & Kapoor, 1986; Reeves, Wagner, Moorman & Kiel 1979; Trappe, 1981, Read, 1991). With ordinary light microscope studies, it is possible to distinguish one or more arbuscular mycorrhizal morphotypes, including the *Glomaceae* and *Gigasporaceae* (Abbott, 1982), the fine-endophyte group (McGonigle & Fitter, 1990) and the *Paris*-morphology group (Widden, 1996).

Background to the Studies

Subject tree species

Four tree species were chosen for study (Table 10.1). Two of these, *Strychnos cocculoides* Baker and *Vangueria infausta* Burch., have been propagated at the VPR nurseries. The first, *S. cocculoides*, has grown very poorly after transplanting into soil in plots at VPR, while the growth of the second, *V. infausta*, has been reasonably good. The two other species chosen for study, *Schinziophyton rautanenii* (Sehiedl) Radel.-sm. and *Bauhinia petersiana* Bolle, are considered by VPR to have a strong potential for acceptance by rural farming communities. Fruits of the *S. cocculoides* and *V. infausta* are routinely sold at the market in Gaborone. In the 1970s, the beans of the *B. petersiana* and the fruits of the *S. rautanenii* were reported to be staple foods for the native Basarwa peoples of the region (Tanaka, 1976; Lee, 1979). The *S. rautanenii* reportedly grows only on Kalahari sand, and it is a protected tree in South Africa (Palgrave, 1977).

Table 10.1 Subject Trees for Mycorrhizal Infection Studies

Identity of tree			Notes of interest about yield
Scientific name	**Family**	**Common name**	
Bauhinia petersiana Bolle	Caesalpiniaceae	white bauhinia	beans contain 25% protein
Schinziophyton rautanenii (Schindl) Radel.-Sm.	Euphorbiaceae	Mongongo	1,000 fruits of 3cm diameter / tree/annum
Strychnos cocculoides Baker	Loganiaceae	wild orange	700 fruits of 10cm diameter / tree/annum
Vangueria infausta Burch.	Rubiaceae	wild medlar	2,000 fruits of 3.5cm diameter / tree/annum roots and pulp are also consumed

Published records reveal that the *S. rautanenii* and the *B. petersiana* form arbuscular mycorrhizae. Other members of the same genus as the *S. cocculoides* have been reported to form arbuscular mycorrhizae (Hogberg & Pierce, 1986; Hogberg, 1982). As well, members of the *Rubiaceae*, the family to which the *V. infausta* belongs, typically form mycorrhizae of the arbuscular type (Harley & Harley, 1987).

Research sites

Samples of the roots of all four study trees and associate vegetation were collected from seven sites in Botswana, southern Zimbabwe, and northern South Africa during early 1997. These sites were visited during one expedition, which left Gaborone on February 25, 1997. The expedition proceeded through Botswana via Francistown, Maun and Shakawe, crossed into Namibia to drive along the Caprivi Strip from west to east, and then passed through Zimbabwe and into South Africa. The last study site, near Pietersburg, South Africa, was visited on March 6, 1997. Prior to this expedition, a small number of samples were taken from cultivated trees at VPR. A second early rainy season sampling of associated vegetation was conducted in Mogorosi, Sese Siding, Manyana, Gabane in Botswana and Wolwegate, Karwrak and Moloto in South Africa in late 1997. Out-planting trials were conducted in Gabane, Shakawe, Plumtree and Pretoria.

Collection of samples

For *S. cocculoides*, *S. rautanenii*, and *B. petersiana* trees, four separate samples of excavated roots were collected from different locations around the base of a single tree specimen at each site. In addition, a set of four cylindrical soil cores, each 7cm in diameter and 15cm in depth, were taken at 90° intervals around the stem or truck. The radial distance from the base of the tree at which the cores were taken was chosen so as to intercept many tree roots. For the associated plant study, a single sample of excavated roots was collected from one individual of each of the plant species found in the immediate vicinity of the study tree. Shoot samples of these associated plants were pressed for drying to assist in botanical identification. Clearing and staining followed the method of Brundrett, Piche & Peterson (1984). Mycorrhizal colonization of roots by arbuscules and vesicles were determined separately, using the method of McGonigle, Miller, Evans, Fairchild & Swan (1990) at x 200 magnification. Qualitative notes on colonization were made while examining roots at x 400 magnification.

Study A Results

Roots and mycorrhizae in the veld

Mycorrhizal colonization of excavated roots of *S. cocculoides*, *S. rautanenii*, and *B. petersiana* was low (Table 10.2). Both *S. cocculoides* and *S. rautanenii* had, at most, 7% of root length colonized in each case. *B. petersiana* had only trace levels of colonization. Within tree species, there were no differences in the extents of colonization for trees in the different age categories.

Table 10.2 Colonization of roots of *S. cocculoides*, *S. rautanenii*, and *B. petersiana* in the field by mycorrhizae. Data are for percentage of root length with arbuscules and vesicles.

Tree species	Site	Tree age class	Sample size	Arbucular colonization		Vesicular colonization	
				Mean	Standard error	Mean	Standard error
				%			
S. cocculoides	Mogorosi	Young	4	7.4	5.7	0.4	0.4
	Rosenfils	Young	4	0.8	0.6	0.0	0.0
		Mature	3	4.3	2.7	1.3	0.7
S. rautanenii	Morosi	Mediate	4	4.4	3.9	8.1	1.9
	Sese	Young	3	0.0	0.0	5.4	1.7
		Mature	3	0.0	0.0	4.0	2.1
	Shakawe	Young	3	4.0	2.2	8.4	4.2
B. petersiana	Tsodilo	Young	3	0.0	0.0	0.4	0.4
		Mature	4	0.4	0.4	2.4	1.4

Mycorrhizal colonization of *V. infausta* roots was found to contrast markedly with that of the roots of the other species examined in this study. Almost all *V. infausta* roots inspected were found to be extensively colonized, with means for the percentage of segments colonized by arbuscules ranging from 78-98% (Table 10.3)

Table 10.3 Colonization by arbuscules and vesicles of the roots of *Vangueria infausta* in the field.

			Roots of V. *infausta*	
Site	Tree age class	Sample size	Arbucular colonization	
			Mean[1] %	Standard Error %
Mogorosi	Young	4	98.1	11.6
	Mature	4	97.8	17.0
Sefhare	Young	4	96.3	9.2
Pietersburg	Young	4	78.0	26.5

[1] means are for percentage of root segments colonized

Discussion

The low colonization of roots of *S. cocculoides*, *S. rautanenii*, and *B. petersiana* were unexpected, given the reported importance of arbuscular mycorrhizae in tropical systems (Janos, 1980). Two interpretations are possible. First, these trees may be functionally non-mycorrhizal in their natural environment, with less than 10% of their root length colonized. The second possibility is that these species form more extensive mycorrhizal associations during periods of more active nutrient acquisition and growth, such as soon after the first rains at the end of the calendar year. The alternative interpretation, that these communities have a general paucity of mycorrhizae at all times, would be inconsistent with available data for most other plant communities (Allen, 1991; Smith & Read, 1996). Further, seasonal patterns for colonization have been reported elsewhere (Brundrett & Kendrick, 1988; McGonigle & Miller, 1993). The occurrence of extensive colonization of the roots of *V. infausta* confirms that the root handling and processing techniques used here were effective for revealing arbuscular mycorrhizae when they are present.

Study B

It is known that mycorrhizality and the plant dependence thereon varies during specific periods in the plant's life cycle, such as during germination and establishment, periods of stress or periods of maximum growth and productivity (Fitter, 1989). Without restrictive effects such as drought and extreme environmental temperatures, mycorrhizae become sufficiently developed in a particular area to provide for the rapid AM infection of new plants in the community (McGee, 1989; Powell, 1977; Read, Koucheki & Hodgson,1976).

However, AM infection in nature cannot be guaranteed (Baylis,1967). Providing the environmental factors are adequate, infection during the critical period of plant establishment could depend largely on the mycorrhizal activity in the associated vegetation near the susceptible plant. Study B explored arbuscular mycorrhizae activity in vegetation associated with the subject trees in their natural habitats. Root density and plant species are among the most critical factors for AM spread by mycelial growth (root to root) (Warner & Mosse, 1982). The hypothesis was that certain species of the vegetation associated with a subject tree would demonstrate an extent of colonisation of AM morphotypes similar to those predominating in the subject tree. This would indicate which plants in the environment could support inoculation of AM infection to encourage the establishment of subject fruit tree seedlings.

The roots of vegetation – mostly grasses, some herbs, shrubs and trees – were collected during two seasons, the dry season collection used in Study A plus a follow-up collection shortly after the first rains. For a variety of reasons, there are seasonal patterns for the occurrence of AM (Gay, Grubb & Hudson, 1982; Land & Schonbeck, 1991; Sanders, 1990). Wetting of arid soils triggers a rapid response of microbial and AM activity (Jacobson, 1997; Nelson & Safir, 1982; Newman, Child & Patrick, 1986). On the basis of these other investigations and the generally low levels of colonisation found in the initial dry-season sampling, it was considered necessary to obtain additional samples after the rainy season began.

Associated vegetation was considered as any vegetation within twice the canopy diameter of the subject tree.

Results

Twenty-six different plant species were found growing in close association with the specimens of the tree species from which samples were taken. These plant species comprised 16 grasses, 4 herbs, 4 shrubs, and 2 trees. The species found most commonly was Lehmann's love grass, *Eragrostis lehmanniana*, which occurred in association with all of the tree specimens studied at Mogorosi and Sese Siding, as well as one of the two specimens studied at Tsodilo. Common fingergrass, *Digitaria eriantha*, was also found frequently, growing alongside all of the tree specimens studied at Mogorosi and close by one of the two specimens studied at Sese Siding. All other species of associated vegetation occurred much less frequently, being found across all sites in association with only one or two of the tree specimens studied. All of the plant species were found to belong to families that typically form arbuscular mycorrhizae (Hogberg, 1982), except for the three species found in the genus *Grewia*. The genus *Grewia* belongs to the family *Tiliaceae*, which has been reported to be ectomycorrhizal (Harley & Harley, 1987).

A t-test was performed on the data to see if seasonal variation was statistically significant for the combined measurements of all plant types for each category of colonisation. Table 10.4 sets out the summary colonization assessment for dry and rainy season samples.

Table 10.4 Summary colonisation percentages for all plants dry and early rainy season

Season	Means (%)	Samples	Standard Deviation
A. Gigasporineae Arbuscular Colonisation			
Dry season	2.15	33	4.1
Rainy season	9.6	38	9.27
B. Glomineae Arbuscular Colonisation			
Dry season	0.81	33	2.72
Rainy season	4.4	38	5.14
C Total Arbuscular Colonisation			
Dry season 2.8	42	5.47	
Rainy season	14	38	11.5
D Non-specific Hyphal Colonisation			
Dry season	68.5	33	17.2
Rainy season	42.5	38	19.7
E Vesicular Colonisation			
Dry season	3.7	42	3.1
Rainy season	4.8	38	6.7

The general trend of higher activity of the arbuscules in the rainy season is readily apparent from Table 10 4. In general, most plants sampled had arbuscule colonisation percentages of less than 10%, but higher levels in the rainy season. But significantly high variation in colonisation figures was found. The occasional plant with extremes of high colonisation figures demonstrate that there at least exists a large capability for AM infection and colonisation in these environments. Even when there is a patchy distribution of the root community and apparent low mycorrhizal activity, some infection was taking place fairly regularly.

Study C

The objective of Study C was to determine for each subject tree species:

- which arbuscular mycorrhiza fungi inoculation type and method is most practical, involving the least input from external sources
- which arbuscular mycorrhiza fungi inoculation type and method is most promotive with respect to early growth and establishment

The hypothesis was that post-transplant arbuscular mycorrhizal infection will promote the growth of young trees being established in semi-arid soils with poor fertility.

In studies conducted on fruit trees in nursery experiments, it has been found that the host tree species determined the success of single versus mixed arbuscular mycorrhiza species inoculum (Hurtado & Sieverding, 1986). Several national institutes in Central and South America involved in nursery fruit tree experiments collectively determined that species soil-fungus combinations with selected arbuscular mycorrhiza fungi were the most beneficial for growth promotion when used as soil inoculum (Sieverding, 1991). There is, however, a paucity of field trials published, and particularly on the post-transplant effects of arbuscular mycorrhizae fungi inoculation.

The following four inoculation methods intended to infect seedlings with arbuscular mycorrhizae at time of outplanting were compared:

- use of a single AM species inoculum (*Glomus claroideum* Schenk & Smith or one of either two species of *Acaulospora* Gerd. & Trappe emend. Berch)
- use of a mixture of the three species above as inoculum
- pre-cropping the outplanting site with Sorghum bicolor L. to raise the inoculum potential of the indigenous arbuscular mycorrhizae in the soils
- seeding an existing natural grass stand for use as an outplanting site to achieve a similar effect

The field trial sites were chosen bearing in mind the most common indigenous fruit tree species naturally occurring in the area. Sites represent varying soil types, in sandveld of deep, well drained soils at Shakawe to southeastern hardveld of very shallow soils at Gabane (De Wit & Nachtergaele, 1990). In each area, two common grasses were chosen as associated vegetation for possible sources of indigenous inocula (Table 10.5).

Table 10.5 Sites and fruit tree species and their associated grass vegetation

Site	Common Fruit Species	Natural Grass Types
Gabane	*S. cocculoides*	*Eragrostis lehmanniana* Nees (Grass-A) *Erochloa mosambicensis* Dandy (Grass-B)
Shakawe	*S. reutanenii*	*Digitaria ciliaris* (Retz) Koeler (Grass-A) *Eragrostis rigidior* Pilg (Grass-B)
Plumtree	*B. petersiana*	*Pogonarthria squarrosa Roem.* & Schult.) Pilg.(Grass-A) *Aristida meridionalis* Henr. (Grass-B)
Pretoria	*V. infausta*	*Pennisetum clandestinum* Chior (Grass-A) *Cynodon dactylon* (L.)(Grass-B)

- **Inoculum preparations:** Three pot cultures of AM fungi were selected from the collection at the University of Pretoria, accession coded 015 (*Acaulospora sp.*, apparently an undescribed taxon), 016 (*Acaulospora sp.*, another apparently undescribed taxon) and 055 *Glomus claroideum* Schenck & Smith. These were each sub-cultured into five "tall pots" (30cm x 10cm), and seeded with *Sorghum bicolor* L. as the host plant. After three months of growth, the shoots were clipped and the root pieces dried in-situ for one month. To prepare the inoculum, the five pots of each culture type were mixed thoroughly to provide a homogenous inoculum of equivocal infectivity throughout. This mixture of soil, infected root pieces, spores and hyphae is termed "whole inoculum". Each culture-type was divided into 45 X 100 gm. sealed plastic packets. A further one and one-half kg of each of the three culture-types was thoroughly mixed and divided into 45 X 100 gm. packets of "mixed species" inoculum. Identification of the fungal inoculum was accurate to the species level in only one of the culture-types. The other two species require further analyses before a positive identification can be made.
- **Ground preparation:** At the end of the 1997 rainy season (April), at each of the chosen sites, an 80-square-metre site area was cleared of all grass and bushes. These cleared patches were kept free from any other plants until November 1997, when outplanting of indigenous fruit tree seedlings was carried out. Grasses were selected at each site, to be used as providers of indigenous AM infection for the outplanted fruit trees. In May 1997, in the areas of the chosen grass species, seeds from the prior season of the same species were sown; the areas were irrigated in an attempt to enhance the inoculum potential of the soils. Concurrently, adjacent to each grass patch, a four-square-metre plot was cleared and pre-cropped with *S. bicolor*, in an effort to enhance local indigenous AM flora in those soil patches. However, the grasses and *S. bicolor* plant growths were not as successful as anticipated. The seeds that were used for seeding or pre- cropping either did not germinate (in grasses), germinated poorly or were eaten by pests (in *S. bicolor*). Three to four attempts of re-seeding did not improve the situation. The only site that had relatively successful stands of *S. bicolor* and grasses was in Pretoria, at the University of Pretoria experimental site.
- **Seedling outplanting and inoculation:** At the beginning of the 1997 rainy season (November), the indigenous fruit tree seedlings raised in sterile media

at VPR were transported and distributed for outplanting at the sites. At each site, seedlings of the same height were chosen to maintain a relative constant, so plant height measurements could be compared at harvest. On the cleared patches at each site, 25 planting holes were dug at 2.5m x 2.5m spacing. Twenty-five 100g packets of inoculum were allocated to the cleared patches. One packet was randomly placed at every planting hole and thoroughly mixed with the soil when filling the holes, in order to provide inoculum to the seedlings. One seedling was then outplanted in each hole. Seedlings planted in holes that did not receive inoculum were considered control plants. Each grass or sorghum plot received four seedlings spaced at least 2.5m apart. All the seedlings were allowed to grow in the field for 3 months. Whenever rains did not fall, the outplanted seedlings were irrigated to maintain adequate moisture levels in the soils. This, however, was poorly managed and inconsistent.

Results

- *Survival and growth promotion:* The survival rates of outplanted seedlings varied greatly between sites, with the highest recorded at Gabane and Rosenfels and lowest at Shakawe with plants at Pretoria showing intermediate survival rates (Table 10. 6).

Table 10.6 Survival of outplanted seedlings at 3 months at various sites in Southern Africa.

Species	**Survival at 3 months** (%)					
Treatment/ species	*S. rautanenii* (Shakawe)	*S. rautanenii* (Gabane)	*S. cocculoides* (Gabane)	*B. petersiana* (Rosenfels)	*V. infausta* (Pretoria)	*V. infausta* (Gabane)
Mix	0	75	100	100	38	88
015	0	50	100	100	50	88
016	0	75	100	100	13	75
055	0	75	75	100	13	75
Control	0	50	100	100	13	88
G- A	0	N/A	N/A	75	50	N/A
G- B	0	N/A	N/A	75	88	N/A
Sorghum1	0	N/A	N/A	75	100	N/A
Sorghum2	0	N/A	N/A	75	88	N/A

Mix = Mixed inoculum of 015, 016 & 055 culture types.
015 = Culture types Acaulospora spp (1)
016 = Culture types Acaulospora spp (2)
055 = Culture type G. claroideum.
N/A = No outplanting done in those treatments
G-A = Grass type A selected per site.
G-B = Grass B selected per site
Sorghum1 = Sorghum plot near G-A.
Sorghum2 = Sorghum plot near G- B

Poor survival of the outplanted seedlings at Shakawe may have been a result of severe drought or heat waves, as the young seedlings were not protected. Similar trends of varied survival were observed between plants of *V. infausta* outplanted at Pretoria and Gabane. At Pretoria, local grass and sorghum plots seem to have promoted higher *V. infausta* survival rates than single or mixed inoculum plots, while at Gabane the mixed inoculum was more successful.

Variance analysis showed significant differences (P=0.05) in growth promotion of seedlings grown in mixed inoculum culture for *B. Petersiana* and *V. infausta* (Table 10.7). The growth rates of the rest of the tree species were not significantly different. The trend, however, was that seedlings grown in mixed inoculum culture treatment consistently attained greater above-ground growth than those in single culture or locally enhanced AM plots. Seedlings in single inoculum culture types (015 & 016) were lower, though comparable to the control at all sites.

Table 10.7 Biovolume of outplanted seedlings at 3 months at various sites in Southern Africa

Species	**Biovolume** (height x diameter)				
Treatment	S. *rautanenii* at Gabane	*S.cocculoides* at Gabane	*B. petersiana* at Rosenfels	*V. infausta* at Pretoria	*V. infausta* at Gabane
Mix	28.6	9.4	9.2*	3.3*	2.7*
015	17.2	5.8	3.1	1.9	2.7*
016	37.7	6.5	4.6	1.2	1.4
055	22.7	6.4	5.1	1.1	1.0
C	19.8	5.1	5.4	2.5	2.0
G- A	NA	0	NA	2.0	NA
G- B	NA	0	NA	1.1	NA
Sorghum[1]	NA	0	NA	2.1	NA
Sorghum[2]	NA	0	NA	1.6	NA

* significant (P=0.05)
Mix = Mixed inoculum of 015, 016 & 055 culture types.
015 = Culture types Acaulospora spp (1)
016 = Culture types Acaulospora spp (2)
055 = Culture type G. claroideum.

G-A = Grass type A selected per site.
G-B = Grass B selected per site.
Sorghum[1] = Sorghum plot near G-A.
Sorghum[2] = Sorghum plot near G- B.
N/A = No outplanting done in those treatments.

- **Root and arbuscular mycorrhizal (AM) infection:** Fine tree seedling roots were collected after ninety (90) days in the field and staining of AM in the roots was done, using the same procedure as in Studies A and B. AM infection of fine roots was low (<30%) in all the seedlings outplanted in all the sites. However, the trend was consistently higher in mixed inoculum than the other three methods at all sites, except at Pretoria, where root sampling was poorly done. The fine roots sampled were inadequate for meaningful staining. Very little infection was found on the few pieces of root examined. The higher levels of infection with mixed inoculum were observed in plants that grew larger in each tree species. In *B. petersiana* and *V infausta*, infection levels were low (<6%) and so was the plant growth. Probably 90 days was not long enough for these *B. petersiana* and *V infausta* species to allow sufficient infection to significantly promote seedling growth at their respective test sites. Infection by indigenous AM occurred in outplanted tree seedlings, as evidenced by infection of control plants. This infection was found at all sites, indicating that indigenous AM flora is active and infects outplanted trees although the growth promotion on seedlings was lower than those that were inoculated. At Pretoria, where *S. bicolor* grew well, the infection levels of AM on *V infausta* were relatively higher (16%) than all other sites (<6%). No correlation was found between biovolume and infection levels in the outplantings.

Discussion

The mixed species inoculum rated high in promoting growth of transplanted seedlings, although significant differences were found only in *V. infausta* and *B. petersiana*. The high seedling mortality in the other species at the other sites could explain why no significant differences were found. The absence of correlation between AM infectivity and growth supports earlier findings (Powell, 1977; Hamel, Dalpe, Furlan & Parent, 1997) that levels of AM infectivity of clover and leek seedlings were only associated with small differences in growth. The low infection levels (<6%) and low growth promotion of single species cultures, compared to the mixed species culture, suggests that the AM infectivity of fine roots in all the species seedlings is enhanced by the presence of multiple AM species. This is as expected in nature, where many species co-exist and colonise specific areas within the root community. Mixed inoculum may therefore have helped establish and promote initial growth in the young seedlings. Outplanting site soil inoculum levels were low. It is therefore apparent that nursery inoculation of seedlings should be encouraged. In addition, an improved selection of inoculum types should be developed to promote optimum AM infection and growth promotion. Where response to *S. bicolor* cultivation was successful, there was a high infection rate in the seedlings outplanted there. This suggests that a cheaper way to introduce AM would be to allow indigenous grasses to grow. But the use of indigenous grasses does not appear to have the potential to completely replace nursery inoculations.

Conclusion

The three studies have raised important questions about the soil-plant relationships of the four native fruit trees. At the time of dry season sampling, the ecosystems in which these trees naturally occur, appeared to have very low quantities of arbuscular mycorrhizal fungi in roots. The rainy season sampling noted an increase in activity. Although colonization was low during the dry season, arbuscular mycorrhizae did indeed occur in the roots of *S. cocculoides* and *S. rautanenii*, giving a base from which they could potentially expand at other times of the year. Colonization of *B. petersiana* was negligible. In contrast, *V. infausta* was heavily colonized with arbuscules. For the establishment of the fruit trees *S. cocculoides*, *S. rautanenii,* and *B. petersiana*, the importance of mycorrhizae remains unclear; its introduction into outplanting sites requires greater numbers of trials. In contrast, *V. infausta* showed high levels of arbuscular colonization. It is tempting to speculate that this mycorrhizal development is what makes it more easily propagated in trials by VPR. The results of the outplanting trials did reveal a more significant role for the mixed mycorrhizae inoculum, and further testing of this approach should be pursued. For *S. rautanenii*, the restricted natural occurrence of this species in Kalahari sand points the way to the testing of appropriate mineral growth substrates for its successful propagation. Given the locally abundant growth of *B. petersiana* on the loose sands of the Tsodilo region, rooting substrate may also prove to be an important consideration for that species.

While the results of these studies have not been conclusive, they have set the stage for further research on the role of mycorrhizae in the establishment of indigenous fruit trees in Eastern and Southern Africa.

References

Abbott, L.K. 1982: Comparative anatomy of vesicular-arbuscular mycorrhizal fungi as formed on subterranean clover. *Australian Journal of Botany* 30: 485-499.

Allen, M.F. 1991: *The ecology of mycorrhizae*. Cambridge University Press, Cambridge.

Baylis, G.T.S. 1967: Experiments on the ecological significance of phycomycetous mycorrhizas. *New Phytol*. 66: 231-243.

Booth, A. 1994: Soils and land use. In M. Chenje and P. Johnson (eds.) *State of the Environment in Southern Africa*. pp.105-132. Harare: Southern African Research and Documentation Centre.

Brundrett, M.C. and Kendrick B. (1988). The mycorrhizal status, root anatomy, and phenology of plants in a sugar maple forest. *Canadian Journal of Botany* 66: 1153-1173.

Brundrett, M.C., Piche, Y., and Peterson, R.L.P. 1984: A new method for observing the morphologies of vesicular-arbuscular mycorrhizae. *Canadian Journal of Botany* 62:2128- 2134.

De Wit, P.V. and Nachtergaele, F.O. 1990: *Explanatory Notes on the Soil Map of the Republic of Botswana* Field Document 30 (AG:BOT85/011), Gaborone, Botswana: Soil Mapping and Advisory Services, FAO 48pp.

Fitter, A.H. 1989: The role and ecological significance of vesicular-arbuscular mycorrhizas in temperate ecosystems. *Agriculture, Ecosystems and Environment* 29: 137-151.

Gay, P.E., Grubb, P.J. and Hudson, H.J. 1982: Seasonal changes in the concentrations of nitrogen, phosphorus and potassium, and in the density of mycorrhiza in biennial and matrix-forming perennial species of closed chalkland turf. *Journal of Ecology* 70: 571- 593

Hamel, C. Dalpe, Y., Furlan, V. and Parent, S. 1997: Indigenous population of arbuscular mycorrhizal fungi and soil aggregate stability are major determinants of leek (*allium porrum* L.) response to inoculation with *Glomus intraradices* Schencnk & Smith or *Glomus versiforme* (Karsten) Berch. *Mycorrhiza* 1977: 7, 187-196

Harley, J.L. and Harley, E.L. 1987: A check-list of mycorrhiza in the British flora. *New Phytologist* 105 (Suppl.): 1-102.

Hayman, D.S. 1983: The physiology of vesicular-arbuscular endomycorrhizal symbiosis. *Canadian Journal of Botany* 61: 944-963.

Hogberg, P. 1982: Mycorrhizal associations in some woodland and forest trees and shrubs in Tanzania. *New Phytologist* 92: 407-415.

Hogberg, P. and Piearce, G.D. 1986: Mycorrhizas in Zambian trees in relation to host taxonomy, vegetation type and successional patterns. *Journal of Ecology* 74: 775-785.

Hurtado, V.M.A. and Sieverding, E. 1996: Estudio del efecto de hongos formadores de micorriza vesiculo arbuscular (MVA) en cinco especies latifoliada reginales en la zona geographica del Valle del Cauca, Columbia, *Suelos Ecuatoriales* 161: pp.109-115.

Jacobson, K.M. 1997: Moisture and substrate stability determine VA-mycorrhizal fungal community distribution and structure in an arid grassland. *Journal of Arid Environments* 35: 59-75

Janos, D.P. 1980: Mycorrhizae influence tropical succession. *Biotropica* 12 (Suppl.) 12: 56-64.

Land, S. And Schonbeck, F. 1991: Influence of different soil types on abundance and seasonal dynamics if vesicular-arbuscular mycorrhizal fungi in arable soils of North Germany. *Mycorrhiza* 1:39-44

Lee, R.B. 1979: *The !Kung San*. Cambridge University Press, Cambridge.

Marx, D.H. 1969: Antagonism of mycorrhizal fungi to root pathogenic fungi and soil bacteria. *Phytopathology* 59: 153-163.

Mateke, S.M. 1998: Field trials to determine the effectiveness of four methods of AM inoculation of out-planted indigenous fruit trees. ECEP Project Report, unpublished.

McCullum, J. 1994: Regional overview. In M. Chenje and P. Johnson (eds.) *State of the Environment in Southern Africa* (edited by) pp.1-20. Harare: Southern African Research and Documentation Centre.

McGee, P.A. 1986: Mycorrhizal associations of plant species in a semi-arid community. *Aust. J. Bot.* 34: 585-593.

McGee, P.A. 1989: Variation in propagule numbers of vesicular-arbuscular mycorrhizal fungi in a semi-arid soil. *Mycol. Res.* 92(1): 28-33.

McGonigle, T.P. and Fitter, A.H. 1988: Ecological consequences of arthropod grazing on VA mycorrhizal fungi. *Proceedings of the Royal Society of Edinburgh* 94B: 25-32.

McGonigle, T.P. and Fitter, A.H. 1990: Ecological specificity of vesicular-arbuscular mycorrhizal associations. *Mycological Research* 94, 120-122.

McGonigle T.P., and Miller, M.H. 1993: Mycorrhizal development and phosphorus absorption in maize under conventional and reduced tillage. *Soil Science Society of America Journal* 57: 1002-1006.

McGonigle, T.P., Miller, M.H., Evans, D.G., Fairchild, G.L. and Swan, J.A. 1990: A new method which gives an objective measure of colonization of roots by vesicular-arbuscular mycorrhizal fungi. *New Phytologist* 115, 495-501.

Mejstrik, V.K. and Cudlin, P. 1983: Mycorrhiza in some plant desert species in Algeria. *Plant Soil* 71: 363-366.

Miller, R.M. 1979: Some occurrences of VAM in natural and disturbed ecosystems. *Can. J. Bot.* 57: 619-623.

Mukerji, K.G. and Kapoor, A. 1986: Occurrence and importance of vesicular-arbuscular mycorrhizal fungi in semi-arid regions of India. *Forest Ecology Management* 16: 117- 126.

Nelson, C.E. and Safir, G.R. 1982: Increased drought tolerance of mycorrhizal onion plants caused by improved phosphorus nutrition. *Planta* 154: 407-413

Newman, E.I. Child, R.D. and Patrick, C.M. 1986: Mycorrhizal infection in grasses of Kenyan savanna. *Journal of Ecology* 74: 1179-1183

Newsham, K.K., Fitter, A.H., and Watkinson, A.R. 1995: Arbuscular mycorrhiza protect an annual grass from root pathogenic fungi in the field. *Journal of Ecology* 83: 991-1000.

Palgrave, K.C. 1977: *Trees of South Africa*. Cape Town: Struik.

Powell, C.L. 1977: Mycorrhizas in hill-country soils I. Spore- bearing mycorrhizal fungi in thirty-seven soils. *N. Zealand. Journal of Agricultural Research*. 20: 53-57.

Read, D.J. 1991: Mycorrhizas in ecosystems. *Experientia* 47: 376-391.

Read, D.J., Koucheki, H.K. and Hodgson, J. 1976: Vesicular-arbuscular mycorrhiza in natural vegetation systems. I. The occurrence of infection. *New Phytol*. 77: 641-653.

Reeves, F.B., Wagner, D., Moorman, T., and Kiel, J. 1979: The role of endomycorrhizas in revegetation practices in the semi-arid West. I. A comparison of incidence of mycorrhizae in severely disturbed vs. natural environments. *Am. J. Bot.* 66: 6-13.

Sanders, I.R. 1990: Seasonal patterns of vesicular-arbuscular mycorrhizal infection in plant root systems. *Plant Soil* 71: 223-246

Sieverding, E. 1991: *Vesicular-arbuscular mycorrhiza management in tropical agrosystems*. Eschborn, Germany: Technical Corporation

Sinclair, R.C. 1998: Arbuscular mycorrhizal activity in vegetation associated with fruit trees in natural habitats. ECEP Project Report, unpublished.

Smith, S.E., and Read, D.J. 1996: *Mycorrhizal symbiosis*. London: Academic Press.

Tanaka, J. 1976: Subsistence ecology of Central Kalahari Sand. In R.B. Lee and I. DeVore (eds.) *Kalahari hunter-gatherers* pp. 98-119. Cambridge, Mass.: Harvard University Press,.

Trappe, J.M. 1981. Mycorrhizae and productivity of arid and semi-arid rangelands. In: *Advances in Food Producing Systems for Arid and Semi-Arid Lands*. New York: Academic Press, pp. 581-589.

Warner, A. and Mosse, B. 1982: Factors affecting the spread of vesicular-arbuscular mycorrhizal fungi in soil. I. Root density. *New Phytol*. 90: 529-536.

Widden, P. 1996: The morphology of vesicular-arbuscular mycorrhizae in *Clintonia borealis* and *Medeola virginiana*. *Canadian Journal of Botany* 74: 679-685.

Zajicek, J.M., Daniels-Hetrick, B.A. and Owensby, C.E. 1986: The influence of soil depth on mycorrhizal colonization of forbs in the tallgrass prairie. *Mycologia* 78: 316-320.9.

CHAPTER 11

INTEGRATED PEST MANAGEMENT IN LOW-INPUT AGRICULTURE IN AFRICA: A REVIEW

by Mark G. Wright

INTRODUCTION

This paper briefly reviews integrated pest management practices used by resource-poor farmers in Africa, considers the problems associated with implementation of pest and disease control for resource-poor farmers in South Africa and briefly discusses some research projects currently being conducted within the South African Agricultural Research Council (ARC).

A BRIEF REVIEW OF IPM IN AFRICA

Integrated pest management (IPM) practices used by resource-poor farmers are essentially similar to those used by any IPM practitioner, except that there may be greater emphasis on cultural control measures such as selection of planting time, crop rotation, sanitation, tillage, intercropping and trap cropping (Poswal, Akba & Alabi, 1993). Early sowing of various crops has been shown to reduce disease incidence such as bacterial pustule (*Xanthomonus phaseoli var sojense*) on soybean (Nweke, 1978), *Cercospera* on eggplant (Tologbonshe, 1983), and rosette virus on groundnuts (Poswal *et al.*, 1993). Crop rotation has been shown to reduce soil-invading organisms such as nematodes, soil fungi and many weeds (Poswal *et al.*, 1993). Planting crops such as maize and cotton together has been shown to increase problems with Lepidoptera accumulating on maize and then moving to late-sown cotton (Onu, 1986). Alternating tomatoes and cereals substantially decreases populations of *Sclerotium rolfsii* in the Nigerian savannah (Wokocha, 1988).

Sanitation is another valuable practice that reduces incidence of pests and diseases. Essentially, sanitation removes sites where pests can hide from predators and reproduction sites. Many pathogens may be controlled by tillage, resulting in the removal of contamination sources. Various soil-borne pathogens (e.g.

Verticillium dahliae, Fusarium oxysporum, Thielaviopsis basicola) can be rendered innocuous by deep ploughing (Poswal *et al.*, 1993). Insect pests (e.g. *Eldana sacharina, Sesamia calamistis, Busseola fusca*) can also be reduced by employing effective tillage (Kaufmann, 1983). Plant intercropping is a traditional practice that has survived among many resource-poor farming communities throughout the world (Norman, 1974) and is a well-know means of reducing pest problems. Intercropping can reduce not only pests, but also diseases. Build-up of anthracnose (*Colletotrichum lindemuthianum*) has, for example, been shown to be slower in mixed crop farms (Williams, 1974).

Trap cropping with dwarf pigeon pea and sun hemp has been found to reduce thrip damage to cowpeas (*Melagurothripos sjostedti*) in Africa and a similar practice using millet and sorghum as a trap crop significantly reduces *Busseola fusca* infestation of millet (Adesiyun, 1983). Planting tall crops like maize around crops susceptible to viral infections has been shown to be effective, as many of the non-persistent aphid-borne viruses are lost before leaving the barrier plants (Kumar, 1984).

Biological control of pests and diseases has not been fully utilized in Africa (Poswal *et al.*, 1993). Some pests have been successfully bio-controlled. For example, the cassava mealybug (*Phenacoccus manihoti*) is controlled in about 24 African countries by a natural enemy (IITA, 1985). Cassava green spider mite (*Mononychellus tanaja*) is also effectively controlled in Benin and Kenya by two predatory mites (IITA, 1990). Biological control of pathogens has met with little success in Africa, owing mainly to inclement weather conditions (Poswal *et. al.*, 1993). The use of resistant cultivars is considered to be valuable to resource-poor farmers, as it is often the most economical alternative available.

Plant extracts are used quite extensively as pesticides in Africa. Neem seed is among the most effective of these remedies, comparable in efficacy to some synthetic pyrethroids (Poswal & Akpa, 1991). Laboratory trials have even shown that neem extract may reduce anthracnose infestations (Poswall *et al.*, 1993). At least 29 other plant species are used as sources of pesticides in other parts of Africa, and a large portion of these are currently being examined carefully by researchers for pesticidal value (Berger, 1994). In Zambia, the use of these botanical pesticides is becoming advanced and research is being conducted there on the possibility of combining these products with sophisticated IPM tools such as pheromone traps (Bahana, 1995). Various traditional remedies for pests and diseases are also used and most of these have not yet been scrutinized for efficacy by researchers.

Problems Associated with the Implementation of IPM in South Africa

Implementing IPM is beset with an intimidating range of problems. The lack of technical knowledge among farmers is perhaps the most critical. This lack of knowledge of IPM philosophy and techniques can be offset by indigenous knowledge within older farming communities. In many areas, mixed cropping and intercropping are practised. The avoidance of pests is one reason. More generally, mixed cropping and intercropping are used to ensure availability of a range of crops and to ensure that at least something may be harvested in the event of the failure of certain crops. However, in many cases, indigenous knowledge of farming systems has become extinct within communities, owing to lack of interest in agriculture. In most cases, extension services in South Africa are poor and little new knowledge filters down to these communities. Resource-poor farmers are, however, receptive to technology transfer and in many cases will implement IPM proposals readily to the best of their ability. It is our experience that farmers attending vegetable production courses presented by the ARC will express interest and request information on intercropping and natural pest control, for example. The lack of technical knowledge can therefore be addressed and alleviated with dedicated training programs and continued effective extension services. A thorough survey of the status of indigenous knowledge of pest and disease control would also be beneficial.

The lack of adequate equipment to deal with pest and disease problems is a substantial problem. Many developing communities can scarcely afford pesticides and the application equipment which should be included in integrated management programs. In such cases, it becomes important to develop relevant technology that directly addresses the needs of the farmers and which can be implemented with limited inputs. This field poses many valuable research problems that may be addressed in a formal scientific manner.

Poor farming conditions add substantially to the problems of pest management. Many resource-poor farming communities in South Africa are relegated to marginal lands that offer little potential for agricultural production. This results in a situation where many crops are cultivated under unsuitable or less than optimal conditions. This makes crops highly susceptible to pests and diseases and, combined with a lack of knowledge and equipment, leads to many failing crops. The implementation of effective IPM under these conditions is difficult and can most probably only be attempted once infrastructure is improved and farmers are provided with relevant information.

Sociological and political problems are often as important as the more technical aspects mentioned above. Many resource-poor farming communities have a healthy distrust for outsiders who appear and attempt to provide information and assistance. The importance of participatory planning in such communities must be emphasized and extension officers and researchers should be aware of this need. Dedicated development programs are also important, because farmers quickly loose impetus and faith if the effort is made only in terms of planting new ideas.

Some IPM Projects within the ARC

The ARC has a large number of projects dedicated to the improvement of IPM for resource-poor farmers.

A project has been undertaken to design a means of selecting crops that will reduce pest incidence when combined. This project examines the effect of intercropping vegetables (and other companion crops) in combinations of crops with similar pests and combinations with dissimilar pests, as well as similar and dissimilar architectures (Wright *et al.*, 1995). The results have shown that combinations of crops with dissimilar pest complexes and dissimilar architectures (e.g. maize, green beans, carrots, sunflower) are most effective. Significantly less produce was lost to pests in these treatments, most likely because these plants harboured a very favourable beneficial insect complex. The ultimate aim of this work is to produce a farmer-friendly guide to selecting suitable crops to plant in combination to avoid pests.

Another aspect enjoying attention is the potential of indigenous plants (specifically *Meliaceae*) as sources of botanical pesticides. This project includes a thorough investigation of these plants in the field and laboratory. The level of interest shown by farmers is also taken into account. So far, a number of plants with pesticidal potential have been identified and farmers are enthusiastic about adopting such technology.

The ability to identify pests and beneficial insects is central to any IPM program. ARC-Roodeplaat is therefore in the process of developing a colour field guide to vegetable pests and their natural enemies. This guide is aimed specifically at the resource-poor farmer. Innovative approaches to help the farmer decide whether an insect is pestilential or beneficial are being devised. The implementation of each of these products needs to be given considerable attention.

Conclusion

IPM is widely used in Africa, particularly the cultural aspects, but there is substantial potential for expansion of on-farm practice and technical knowledge. Diffusion of basic information on pests and diseases is important. But it must be recognized that IPM programmes develop in stages and are not ready-to-use. Ongoing interaction between researchers and farmers is critically important to the design of IPM strategies that fit well into existing farming systems. Thus a participatory and interdisciplinary approach to IPM development will encourage adoption by farmers. Support for ongoing training and research is of cardinal importance.

References

Adesiyun, A.A. 1983: Some effects of intercropping of sorghum, millet and maize on infestation by lepidopterous stalk-borers, particularly *Busseola fusca*. *Insect Science & Application*, 4, 387-391.

Bahana, J.W. 1995: The use of pheromone traps to monitor the African armyworm, *Spodoptera exempta*, Walker, in Zambia and the possibility of timely application of natural products to control outbreaks. In *Natural Plant Products as Pesticides*, pp. 110-114.

Berger, A. 1994: Using natural pesticides: current and future perspectives. *Swedish Univ. Agric. Sci.*

IITA (International Institute for Tropical Agriculture). 1985: *Research highlights for 1984*. Ibadan, Nigeria.

IITA. 1990: *Annual Report*. Ibajan, Nigeria.

Kaufmann, T. 1983: The behavioral biology, feeding habits and ecology of three species of maize stem borers: *Eldana sacharina* (Lep: Pyralidae), *Sesamia calamis* and *Busseola fusca* (Noctuidae) in Ibadan, Nigeria. *Journal of the Georgia Entomological Society*.

Kumar, R. 1984: Physical control of pests. In *Insect pest control – with reference to African Agriculture*, pp. 63-71. U.K.: Camelot Press.

Norman, D.W. 1974: Rationalizing mixed cropping under indigenous conditions. The example of northern Nigeria. *Journal of Developmental Studies*, 11, 3-21.

Nweke, M.E. 1978: *Studies on the bacterial pustule disease of soybean in northern Nigeria*. M.Sc. Thesis. Nigeria: Ahmadu Bello University.

Onu, I. 1986: Present status of cotton insect pests in the northern Guinea savanna of Nigeria. 16th Annual Conference of the Nigerian Society for Plant Protection.

Poswal, M.A.T. and Akpa, A.D. 1991: Current trends in the use of traditional and organic methods for the control of crop insect-pests and diseases in Nigeria. *Tropical Pest Management*, 37, 329-333.

Poswal, M.A.T., Akpa, A.D. and Alabi, O. 1993: Cultural control of pests and diseases: Prelude to integrated pest management practices for resource-poor farmers in Nigerian agriculture. *Journal of Sustainable Agriculture,* 3, 5-48.

Tologbonshe, E.D. 1983: *Studies of Cercospera leaf spot disease of egg plant* (Solanum spp.) *in the Zaria area.* M.Sc. Thesis. Nigeria: Ahmadu Bello University.

Williams, R.J. 1974: Plant pathology in the grain legume improvement programme at IITA. Proceeding of the First Grain Legumes Workshop IITA, Ibadan.

Wokocha, R.C. 1988: Relationship between the population of viable sclerotia of *Sclerotium rolfsii* in soil to cropping sequence in the Nigerian savanna. *Plant and Soil*, 106, 146-148.

Wright, M.G., Rencken, I.C., Claasen, E.M. and Hattingh, V. 1995: Selecting vegetables to intercrop as a pest avoidance strategy: preliminary results. Proceedings: Tenth Entomological Congress, Entomological Society of Southern Africa. Grahamstown, S.A.

CHAPTER 12

THE USE OF NEEM FOR BOTANICAL PEST MANAGEMENT

by Ramesh C. Saxena

INTRODUCTION

It is now well recognized that sustainable agricultural systems must be ecologically sound for long-term food sufficiency, equitable in providing social justice, and ethical in respecting both future generations and other species. Sustainable pest management is relevant to farming in semi-arid environments of Africa fraught with economic risks and uncertainties due to a wide range of unpredictable biotic and abiotic stresses. Cash-strapped, resource-poor farmers have little or no means to purchase expensive pesticides. Even if synthetic pesticides could be made available at subsidized prices, the overall benefits resulting from their use may not be worth the risks to public health and the environment. Inexpensive, relatively safe and environmentally friendly pest management technologies are therefore needed to enhance agricultural productivity of staple food crops.

The International Centre of Insect Physiology and Ecology (ICIPE) is an internationally known institute devoted to alleviating poverty and ensuring food security and good health for the peoples of the tropics through the management of both harmful and useful arthropods. It operates three focal programmes: Plant Pest Management, Disease Vector Management and Arthropod Biodiversity Conservation and Utilisation. The Plant Pest Management Programme emphasizes the reduction of pest infestations through biological control, habitat management, botanical pest control, and the deployment of pest-tolerant cultivars. As an example of ICIPE's research on botanical pest control this paper will focus on the use of neem for stem borer management in maize and sorghum, which are important food crops in sub-Saharan Africa. The lepidopterous stem borers seriously limit the potential of attainable maize and sorghum yields. Even at a conservative estimate, 20-40% of the African crop yield is lost annually to stem borers, the most important being *Busseola fusca* and *Chilo partellus*. The native stem borer *B. fusca* occurs throughout sub-Saharan Africa, while *C. partellus* is an exotic species accidentally introduced into Africa from Asia in the early 20th century and currently restricted to Eastern and Southern Africa.

Neem

In recent years, there has been growing interest in the use of pesticidal plants, especially the neem tree *Azadirachta indica* (A. Juss.), already known for its pest control properties. Native to the drier parts of the Indian sub-continent, neem is now widely distributed in Africa. Every part of the tree is bitter and has pest control properties, but maximum concentrations of highly bio-active limonoids, including azadirachtins, are found in the seed. Azadirachtins are structurally complex molecules which render their synthetic production uneconomic. The use of simple formulations or mixtures of bioactive principals occurring naturally in the neem seed would be highly desirable and sustainable for pest management on African farms. Though subtle, neem's effects include:

- repellence
- feeding and oviposition deterrence
- disturbing insect host-endosymbiote equilibrium
- growth inhibition
- mating disruption
- chemosterilization
- reduced egg hatchability

These are far more desirable than a quick knock-down effect because they reduce the risk of exposing pests' natural enemies to either poisoned food or starvation (Saxena, 1989; Schmutterer, 1990, 1995; National Research Council, 1992; Ascher, 1993). In spite of high selectivity, neem derivatives affect more than 400 species of insect belonging to various groups of insects. The lepidopterous pests are particularly sensitive to neem.

Results from Neem Trials

In recent years, the results of our investigations using neem materials against common species of maize and sorghum borers seem to be highly encouraging. In trials conducted at the ICIPE Field Station and in farmers' fields at Mbita in Kenya, powdered neem seed was applied to maize which had been infested with eggs of the spotted stem borer, *Chilo partellus*. The foliar application of neem significantly reduced the foliar damage, stem tunnelling, tassel breakage and populations of the stalk borer larvae (ICIPE, 1993, 1994, 1995). Grain yield in neem-treated maize plots was as high as that obtained with insecticides and significantly higher than that in untreated control plots (Table 12.1).

Table 12.1 Comparative damage by *Chilo partellus* larvae as measured by tassel breakage and grain yield in plots planted to 'Katumani' a maize cultivar susceptible to stem borer. Data from the short rains cropping season 1992[1]

	Experiment Station		Farmer's Field	
Treatment	**Tassel breakage (%)**	**Yield (kg/ha)**	**Tassel breakage (%)**	**Yield (kg/ha)**
NSP[2] (basal)	17.0 b	4530 b	12.0 b	3570 b
NSP (foliar)	2.0 a	6430 a	4.0 a	5480 a
NSP (basal+foliar)	2.0 a	5870 a	2.0 a	5630 a
Furadan 5G	0.3 a	6400 a	0.3 a	6130 a
Untreated (control)	21.0 b	3370 b	17.0c 3850 b	

1 Within each column means followed by the same letter are not significantly different at the 5% level by the Least Significant Difference test; averages of 4 replications
2 NSP = neem seed powder

Late infestation of maize by field populations of *Eldana saccharina* was low in neem- treated plots, but not in insecticide-treated or untreated control plots. Storage of neem cake up to two years in the dark did not reduce its effectiveness against stem borers (Table 12.2).

Similar reductions in pest damage, including body size as measured by the width of the head capsules, and increases in grain yield were obtained when neem cake was applied to the sorghum crop, both at the Experiment Station and in a farmer's field (Table 12.3). Because the cost of neem treatment was approximately one-tenth of the cost of the insecticide treatment, the net gain was higher with neem.

Table 12.2 Comparative stem borer infestation and plant damage by *Chilo partellus* and *Eldana saccharina* larvae and grain yield in plots planted to 'ICZ5', a moderately resistant maize cultivar and applied with neem cake (NC) or a synthetic insecticide. ICIPE Experiment Station, Mbita, Kenya. Long rains cropping Season 1994.[1]

Treatment	Damage[2]	Plant height (cm)	Tunnel length (cm)		Chilo (no.)	Eldana (no.)	Tassel Breakage (%)	Yield (kg/ha)
	9 WE[3]	14 WE	7 WE	14 WE	9 WE	11WE	11 WE	
Fresh NC	2.4 a	180ab	0.6 a	15.6 a	1.5 a	7.8 a	8.6 a	7458 a
1-yr-old NC	2.4 a	186 a	0.7 a	13.0 a	1.5 a	6.5 a	8.1 a	7760 a
2-yr-old NC	2.4 a	185 a	0.6 a	18.5 a	0.3 a	9.4 a	9.4 a	7469 a
Dipterex	2.1 a	177ab	0.5 a	20.3ab	0.5 a	17.0 a	17.0 a	7271 a
Control (untreated)	4.6 b	167 b	7.8 b	31.6 b	14.3 b	25.5 b	35.5 b	5088 b

1 Within each column means followed by the same letter are not significantly different at the 5% level by the Least Significant Difference test; averages of 4 replications

2 Foliar damage scored visually on a 1-9 scale (1 = no damage, 9 = completely damaged)

3 WE = weeks after crop emergence

Table 12.3 Comparative plant damage, *Chilo partellus* larval infestation and grain yield in plots planted to 'Serena', a susceptible sorghum cultivar and applied with neem cake (NC) once at 4 WE[1] or twice at 4 and 6 WE or with Dipterex. ICIPE Experiment Station and farmer's field, Mbita, Kenya. Short rains cropping Season 1994.[2]

Treatment	Damage[3] at 9 WE	Plant height (cm)	Tunnel length (cm)	Larvae (no.) at 9 WE	Head width (mm)	Yield (kg/ha)
Experiment Station						
NC once	2.9ab	110 a	27.5ab	7.4ab	0.66 a	6182ab
NC twice	2.9ab	106ab	23.1 a	2.2 a	0.67 a	7312 a
Dipterex	2.3ab	117 a	20.3 a	7.0ab	0.79ab	6523ab
Control (untreated)	3.5 b	94 b	30.3 b	10.2 b	0.84 b	6056 b
Farmer's Field						
Treatment	Damage[3] at 9 WE	Plant height (cm)	Tunnel length (cm)	Larvae (no.) at 9 WE	Head width (mm)	Yield (kg/ha)
NC once	2.9 a	114 a	25.4 b	18.8ab	0.64	5242 a
NC twice	3.0 a	112 a	18.9 a	16.8 a	0.63	5052 a
Dipterex	2.5 a	111 a	22.9ab	19.4ab	0.96	5043 a
Control (untreated)	4.1 b	99 b	27.9 b	25.8 b	0.90	3908 b

1 WE = weeks after crop emergence
2 Within each column means followed by the same letter are not significantly different at the 5% level by the Least Significant Difference test; averages of 4 replications
3 Foliar damage scored visually on a 1-9 scale (1 = no damage, 9 = completely damaged)

We also evaluated the effectiveness of neem cake in plots in a farmer's field artificially infested with the maize stem borer, *Busseola fusca*. As the level of infestation achieved was not high enough and as *B. fusca* is a slow feeder, the differences in foliar damage and plant heights were not significant in treatments with neem cake applied once or twice, Dipterex or in untreated control. Nonetheless, neem cake or insecticide-treated plants had significantly less stem tunnelling and fewer surviving *B. fusca* larvae than the control plants (ICIPE, 1994).

To determine the cause of poor establishment and low plant damage by the stem borer *C. partellus* on neem-treated maize and sorghum plants, we studied the growth and development of neonate larvae on a standardized diet containing 25, 50, and 100 parts per million of neem seed powder prepared from freshly harvested seed. We found that growth and development of larvae were significantly reduced on the diet containing >50 ppm of neem seed powder (ICIPE, 1994).

Exposure of larvae to a diet containing neem seed powder also reduced the fecundity of emerged females and the hatchability of their eggs. To determine the cause of poor establishment of *B. fusca* on neem-treated maize plants, we studied the growth and development of *B. fusca* neonate larvae on a semi-synthetic diet to which 10, 100, 500, or 1000 parts per billion of "neem bitters concentrate" containing 10% azadirachtin was added. Larval growth was significantly reduced and the developmental period prolonged on the diet containing neem bitters concentrate at >100 ppb; larval growth was totally inhibited at >500 ppb (ICIPE, 1995). The fecundity of females developing on the diet with neem bitter concentrate at 100 ppb was significantly reduced, as was the hatchability of eggs.

Other meliaceous plants, such as the Persian lilac or chinaberry tree, *Melia azedarach L.* and *Melia volkensii Gurke*, which are widespread in cooler parts of Africa, also have pest control properties (Schmutterer, 1995). Their chemical aspects have been reviewed in depth (Kraus *et al.*, 1993, Rembold & Mwanigi, 1995). Toosandanin is the major bioactive material in the bark of *M. azedarach*, while meliantriol and melianone occur in fruits of *M. azedarach*. Volkensin occurs in the fruit extract of *M. volkensii*. But it is difficult to process the hard Melia seed and the presence of toxic meliacins in Melia fruits and bark limits their use as insecticides.

Conclusion

The use of botanicals for pest control originated as a need and not science in many parts of the world in ancient times, but with the development of better techniques and analytical tools for detection, isolation, and identification of bioactive constituents, the scientific basis of botanical pest control has now emerged on a firm footing. However, for the sake of durability, botanicals such as neem should still be used within the framework of an integrated pest management strategy. In this context, the relative safety of neem materials to beneficial organisms, including predators, parasitoids, and crop pollinators could confer an added advantage (Schmutterer, 1995).

References

Ascher, K.R.S. 1993: Nonconventional insecticidal effects of pesticides available from the neem tree, *Azadirachia indica. Arch. Insect Biochem. PhysioL*, 22, 433-449.

ICIPE (International Centre of Insect Physiology and Ecology). 1993: *Annual Report Highlights.*

ICIPE (International Centre of Insect Physiology and Ecology). 1994: *Annual Scientific Report.*

ICIPE (International Centre of Insect Physiology and Ecology). 1995: *Annual Report.*

Kraus, W., Bokel, M., Schwinger, M., Vogier, B., Soeliner, R., Wendisch, D., Steffens, R. and Wachendorff, U. 1993: The chemistry of azadirachtin and other insecticidal constituents of *Meliaceae*. In T.A. Van Beek and H. Breteller (eds.), Phylochemistry and Agriculture. Proceedings of the *Phylochemistry Society of Europe*, pp. 18-39. Oxford Univ Press.

National Research Council. 1992: *Neem - A Tree for Solving Global Problems.* Report of an ad hoc Panel of the Board on Science and Technology for International Development. Washington: National Academy Press.

Rembold, H. and Mwamigi, R.W. 1995: *Melia volkensii* Gurke. In H. Schmutterer (ed.), *The Neem Tree, Azadirachta indica A. Juss.: Source of Natural Products for Integrated Pest Management, Medicine, Industry and Other Purposes*, pp. 647-652. Weinheim: VCH.

Saxena, R.C. 1989: Insecticides from neem. In J.T. Arnason, B.J.R. Philogene and P. Morand (eds.), *Insecticides of Plant Origin. A CS Symposium Series* 387, pp. 110-135. Washington: American Chemical Society.

Schmutterer, H. 1990: Properties and potential of natural pesticides from the neem tree, *Azadirachta indica. Annu. Rev. Entomol*, 3 5, 271-297.

Schmutterer, H. (ed.). 1995: *The Neem Tree, Azadirachta indica A. Juss.: Source of Natural Products for Integrated Pest Management, Medicine, Industry and Other Purposes.* Weinheim: VCH.

CHAPTER 13

CONTROLLING NEMATODES ON HORTICULTURAL CROPS: EXPERIMENTATION AND TRAINING IN NAMIBIA

by Sarah Rosloski, Peter G. Kevan and John W. Potter

INTRODUCTION

Drought affects many areas of Namibia on an unpredictable and recurrent basis. The Kavango is situated on the northeastern border of Namibia. Forty percent of the residents practice subsistence agriculture (World Bank, 1992) and the economy has been traditionally communal. Fifty percent of households are food insecure. Because the Kavango region consistently has one of the highest rainfall accumulations in Namibia one of the first items on the post-apartheid agenda for development was the improvement of the area's food production capacity (Yaron, Janssen & Maamberua, 1992). OXFAM Canada began to work in the Kavango in 1990, just after independence. In an attempt to alleviate poverty and promote food security, OXFAM sought to encourage and facilitate advances in the agricultural sector. OXFAM, through their subsidiary organization CANAMCO, established programmes which attempted to improve the yields of traditional crops such as pearl millet. To enable farmers to harvest with less manual labour, they also attempted to introduce agricultural technology, such as animal traction devices and threshing and milling machines. CANAMCO endeavoured to introduce horticultural plants (e.g. carrot, cabbage, tomato, onion, squash) and the skills to manage these crops in small-scale production gardens. Because these crops have a high market value, this aspect of the CANAMCO project contributed both to both food and income security. But if new horticultural crops are to be introduced in the Kavango, then solutions to their associated pest problems must also be available. The activities reported in this paper were directly connected to the horticultural aspect of CANAMCO's agricultural work in the region. A research project and a training course were conducted at the Mashare Agricultural and Rural Development Institute. The research proceeded from May 1, 1996 to August 31, 1997. It explored the control of parasitic nematodes in tomato production, using the marigold *Tagetes patula* and the Mexican sunflower *Tithonia diversifolia*. The associated course on nematode management was held from July 28 to August 1, 1997.

The Root-Knot Nematode

Introduced horticultural crops have a high burden of pest pressure in the Kavango region. This is thought to be a direct result of the savanna environment ecology which encourages a top-heavy herbivore density in the food chain (Swift, Heal & Anderson, 1979), and an indirect result of the lack of inbred resistance to these pests in introduced crops. Some of the most serious pests in this region and in Africa, as well as in similar places throughout the tropical and subtropical world, are plant parasitic nematodes, particularly *Meloidogyne spp.*, the root-knot nematode (Luc, Siroka & Bridge, 1990). Root-knot nematodes are obligate root parasites that feed on the roots of many plant species (Agrios, 1997). During infection, the nematode enters the root and induces, through plant cell division, the formation of a large cluster of cells within which the nematode is protected. This cluster appears as a root gall to the observer (Fig. 13.1).

Figure 13.1 Tomato plant infested with nematodes (photo S. Rosloski)

The nematode also stimulates the formation of a large cell, upon which it feeds. This giant cell, or coenocyte, gives the infesting nematode the ability to divert nutrients made by the plant and intended for plant growth. The diverted nutrients support the nematode and its reproductive efforts (Figure 13.2).

Figure 13.2 Nematode parasite inside a root.

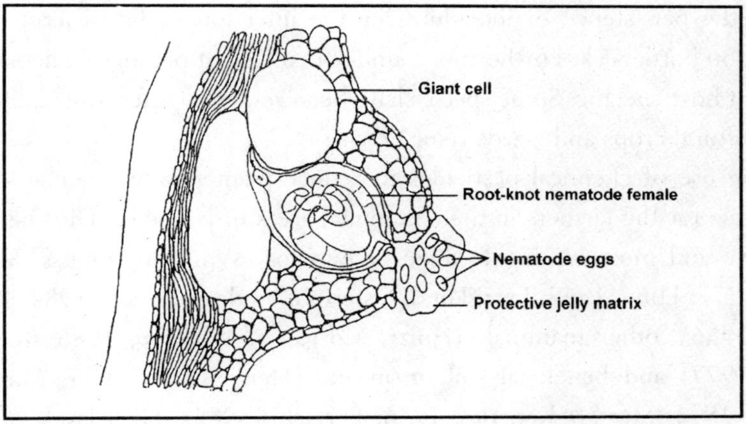

These coenobial cells anatomically resemble transfer cells. Transfer cells in physiologically normal plants act as cellular locations of nutrient dispersal from the phloem (vascular) tissue to tissues in need of food. Thus, the nematode induces transfer-like cells that attract nutrients to themselves. The nematode parasitism reduces the host plant's uptake of water and nutrients because the galls become disruptive to the conduit systems in and out of the root (Figure 13.3).

Figure 13.3 A healthy versus a nematode-infected plant

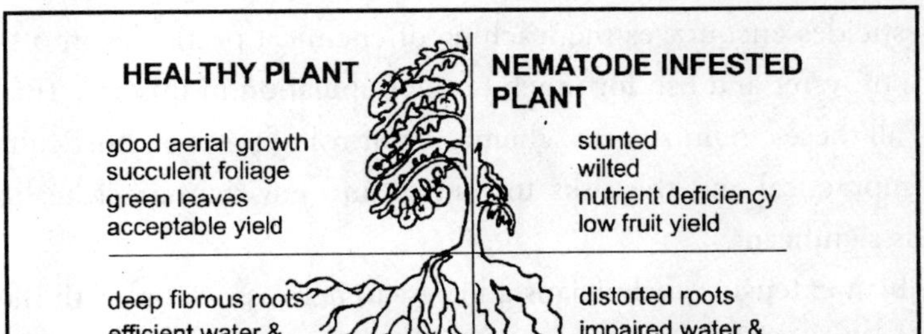

Management of plant parasitic nematodes is a global problem. Damages from these agricultural pests are estimated to be 10-20% of crop yield worldwide. The burden of disease is unequally distributed to tropical and sub-tropical areas of the globe, where high temperatures and low soil fertility favour the heaviest infestations. In tropical and sub-tropical regions, this loss estimate may rise to 60% of crop yield (Luc *et al*., 1990). These parasites build up quickly in virgin soils and the lengthy persistence of inoculum for the infection of future crops causes a production bottleneck. Furthermore, unlike many plant pathogens, nematodes are often not host-specific. Some species have been shown to infest over 2,000 species of agricultural crops and weeds (Noe, 1988).

The use of chemical pesticides to reduce plant parasitic nematodes is not sustainable for the farmers in the communal areas of Namibia. The chemicals are expensive and most are highly toxic to humans (Whorton, Krauss, Marshall & Milby, 1977; Hine, 1969; Letz, Pond, Osterloh, Wade & Becker, 1984; Behrens & Duke, 1986), other mammals (Hurtt, Morgan & Working, 1986; Torkelson & Oyen, 1977) and beneficial soil organisms (Menge, 1982; Ohr, Munnecke & Bricker, 1973; Starr & Kenerley, 1986; Bartschi, 1982). High levels of illiteracy often preclude adherence to dosage or safety precautions found on product labels, an absolute necessity when using chemicals of high toxicity (Bull, 1982; World Bank, 1992). Chemical means of pest reduction also rely on the farmers having investment capital to purchase pesticides. In addition, vegetables are grown along the Kavango River bank because of the requirement for irrigation, the shortage of rainfall through much of the year, the lack of water distribution infrastructure (Yaron *et. al*., 1992), and the inadvisability of using groundwater for irrigation (World Bank, 1992). This use of river water and the inability of sandy soils to bind residual pesticides encourages the leaching of chemical pesticides into the river, a vital source of water and fish for much of the population in this area (Bird, 1987). Because of all these circumstances, chemical control of pests on horticultural crops is largely impractical and the risks to human and environmental health through poisoning is significant.

Namibian extension technicians and researchers associated with the research programme and involved in the course already had an understanding of the symptoms of pest or disease outbreaks that commonly affect crops in the region and the implementation of conventional control strategies for these pests. Both the research programme and the course attempted to provide the basis for pest management alternatives which use modified growing practices and low input or indigenous resources. The research explored the biological control of nematode root pathogens, while the course was used for generalized discussion on various environmentally sensitive techniques for pest control.

The Research

There are several alternative methods of nematode control which do not rely on the use of chemicals. Our first goal was to test two such methods for the control of root parasitic nematodes afflicting tomatoes in the region. These were cover cropping and green manuring. We wanted to determine the efficacy of the techniques in the area and to demonstrate our results to farmers. The work was done at one of the region's agricultural development centres, Mashare Agricultural and Rural Development Institute, about 60 km east of Rundu.

Cover cropping experiment

In the cover cropping experiment, we hypothesized that nematode root parasites of tomato could be controlled by the use of two potentially nematode-suppressive crops, marigold (*Tagetes patula*) and the Mexican sunflower (*Tithonia diversifolia*). The hypothesis was tested in an agricultural field experiment with two seasons of growth and using an eight treatment by five replicate design. The treatments consisted of three plots of the sunflower at densities of 40 cm, 50 cm and 60 cm, three plots of marigold at the same densities, a clean fallow control and a practical check plot of tomato. Prior to planting the marigold and sunflower treatment plants, initial populations of plant parasites were sampled to determine the natural levels available for infection. After removal of the treatment plants, the soil was again sampled to determine if either of the plants were host to any parasitic nematode species. Populations of soil-borne worms were measured to assess whether or not the treatments conferred any effect on subsequent levels of tomato infestation. Tomatoes were then planted in the treated soil and the effects of each treatment on tomato growth were to be assessed.

Results: Neither the marigold nor the sunflower supported reproduction of the root-knot nematode. Root-knot nematode was observed in tomato plots grown adjacent to the two treatment plots, but not in the treatment plots themselves. We conclude that conditions for the worm's reproduction were thus otherwise satisfied but that the marigold and sunflower cover crop plants did not act as a host.

This conclusion is not definitive, however, because the population of worms in the study plots was low at the onset of the experiment. Re-testing in heavily infested soils is required. It was also observed that the two cover cropping plants supported other nematode genera in the soil. At high planting densities, both plants showed receptivity to the nematode genus *Helicotylenchus*. It is not known whether this worm has the potential to be pestiferous on tomatoes or other crops. This observation points to the importance of further study of nematode problems in horticultural production in the Kavango. Tropical and sub-tropical areas may be plagued by multi-species infestations of nematodes. Understanding how each species affects crop vigour and how they interact to produce disease are needed. Ultimately, such ecological knowledge will produce recommendations that will prevent one crop supporting a nematode fauna inimical to the yield of the next crop. The results of the cover cropping experiment are inconclusive with respect to practical nematode control for tomatoes, because the tomato plants used to assess the effect of the treatments were killed by an early frost soon after transplanting.

Green manure experiment

The hypothesis for the green manure experiment was that nematode root parasites of tomato could be controlled by the incorporation of vegetable residues (stems and leaves) of the sunflower into the soil as a green manure. Also, it was postulated that the effects of green manure as a suppressant of nematode populations might be enhanced by the combination of green manure and host resistance in the crop plant. Resistance to the root-knot nematode is available commercially through seed with the resistance gene, called Mi. It is currently the only form of resistance to this nematode known in tomato, but it has several drawbacks. It is conferred by one gene and thus it is potentially simple for the nematode to develop resistance to it. As Mi is unstable at temperatures higher than 28°F, it can loose efficacy in hot climates where it is most needed. The cultivars that have this resistance are also much more expensive than other seed and not affordable for resource-poor farmers.

The efficacy of green manure to control plant parasitic nematodes was compared using a resistant tomato cultivar (cv. STAR 9001) and a susceptible tomato cultivar (Flora-dade). Again, the hypothesis was tested with an eight treatment by five replicate field experiment. The treatments consisted of three plots of susceptible genotypes of tomato with applications of green manure at rates of 50 kg, 100 kg and 150 kg per 19.25 m^2 plot, three plots of resistant genotypes of tomato with the same rates of green manure, a control plot with only resistant plants and no green manure added, and a control plot with susceptible plants with no green manure added. Estimates were made of initial populations of rhizosphere-dwelling plant parasitic worms and final populations, after the growth of the tomato plants. Plant health characteristics, such as leaf area, and fortnightly-measured plant heights and survivorship within plots were recorded.

Results: Green manuring increased many indicators of plant health. Leaf area, stem weight, fruit weight, plant height and survivorship all increased in comparison to the controls. The increase of weight, size and survivorship of the tomato plants is likely attributable to the soil improvement from manuring. Green manure probably increases the amount of nutrients available for plant growth and the aeration or water retention properties of the soil. Both cultivars responded similarly to green manuring, in terms of most indicators of plant health. However, the fruit weight of the nematode susceptible cultivar was higher than that of the resistant cultivar. This likely reflects the susceptible variety's better performance in soils which are not heavily infested, as in our experiment. Resistance usually carries a metabolic cost to the plant resulting in a lower yield. However, in highly infested soils, this lower yield characteristic would be offset by protection resulting in a higher yield than non-resistant cultivars. The resistant cultivar showed increased survivorship compared to the susceptible one.

Beyond nematodes, there are many other pests associated with tomatoes in Kavango. Our tomatoes suffered from infestations of aphids and a stem-girdling, fungal disease. Several plants (approximately 1%) became stunted and chlorotic possibly because of aphid transmitted virus. The different cultivars likely had different responses to these pressures, which increased the survivorship of the resistant plant in comparison to the susceptible plants. Despite the higher cost of seeds, increased survivorship in the face of multiple selection pressures is a strong positive feature of Star 9001 for subsistence farmers who lack capital to buy pest-control products.

Even though green manuring improved plant health, the populations of soil nematodes were not significantly different from those in control plots (Table 13.1). Significant differences were observed between plots with the non-resistant cultivars and the plots with Star 9001. Star 9001 had the greatest reduction of soil-borne root-knot nematodes. In fact, almost no root-knot nematodes were found in the soil taken from the root zone of the Star 9001 cultivar. It has been occasionally observed in previous studies that races of root-knot nematode able to infect otherwise resistant tomatoes may be found prior to the use of these cultivars or that resistance-breaking races of nematodes may evolve after prolonged use of the resistant cultivars. It can be concluded that resistance-breaking races of nematodes are not now present in the soils of Mashare. The use of resistant cultivars was the most effective method of nematode control studied in this project.

Table 13.1 Final populations of *Meloidogyne* ssp. in the green manure experiment

	Nematodes per 300 gm soil sample							
	Sensitive Cultivar				Resistant Cultivar (Star 9001)			
	Manure cover by weight/19.25 m²				Manure cover by weight/19.25 m²			
Sample#	0kg	50kg	100kg	150kg	0kg	50kg	100Kg	150Kg
1	0	7	10	15	0	0	0	0
2	2	2	1	2	0	0	0	0
3	41	2	155	0	0	0	0	0
4	120	12	34	0	0	0	0	1
5	12	0	0	1	0	0	0	0
Mean	35	5	40	4	0	0	0	0

The lack of significant differences in nematode populations between susceptible control plots and susceptible treated plots should not exclude green manuring from the research agenda in the Kavango region. The experiment may require more replications to remove the effects of variability often seen in nematological experiments. Both the 50 kg and 150 kg plots showed reductions in the average populations of root-knot nematode. Also, the sunflower manure had the unfortunate ability to root upon cutting and internment, which may have rendered a portion of the treatment unavailable for decomposition. The efficacy of the treatment is thought to depend on the decomposition of the amendment and release of the products of decomposition to the soil.

Future work with the Mexican sunflower should shred the manure into small, quickly decomposing pieces. Use of a machete to shred the sunflower is incompatible with this goal and other methods of manure preparation should be sought. Alternatively, other species of green manure that do not root may be sought. We recommend several species of plant found in the Kavango be assayed for nematode control when used as green manure, i.e. Castor bean (*Ricinus communis*) and Jimson-weed (*Datura stramonium*). Farmers may welcome the use of those species, because they are weeds. Green manuring should be explored in greater depth because it is one of the few nematode control techniques that may be cost-effective for Kavango farmers and it does not necessitate using chemicals that are highly toxic to people and the environment.

The Course

The theme of the course was that pests can be controlled through an understanding of the ecology of a pest in the agricultural environment and through the introduction of biological mortality factors or biological control. This course theme reflects practical application of 'AgrECOLture' (Kevan, Thomas & Belaoussoff, 1997) as a foundation of a programme of integrated pest management. IPM can be defined as the suppression of a pest below an economic threshold, using several strategies of control at once or in a schedule of rotation. The concept of IPM came into being with the realization that the agricultural forum is really a biological forum. The agroecosystem behaves no differently than a natural system, except that it has been excessively simplified.

Despite the simplification of the biological processes, the operations of natural selection and survival of the fittest still hold dominion as they do in other ecosystems. One rude awakening that has forced agronomists to recognize this attribute in agriculture has been the development of resistance of many pests to pesticides (Georghiou & Taylor, 1977; Luck, van den Bosche & Garcia, 1977). Under simple selection pressures such as a mortality factor (e.g. pesticide), most and even almost all individuals in a population may be killed. However, some individuals persist because of genetically based resistance mechanisms. If a given pesticide is used continuously, the initially small population of resistant pests eventually increases to the levels before control was instituted. If a pesticide has a simple mechanism of action, a pest has only to possess a simple genetic attribute to overcome the chemical. It follows that the more diverse or complex the selection pressure (treatment), the less likely that variants that can resist the treatment will be present. The more diverse the pest-control strategy, the less likely the possibility that any simple selection can occur.

Integrated Pest Management does not preclude the use of synthetic chemical pesticides, but includes them with other means of pest control such as biological, cultural, and sanitary techniques (Campbell, 1989). IPM practitioners realize that pesticides, used as a sole method of control without detailed knowledge of the most advantageous application times or rates, are a tactic that can quickly become obsolete through the processes of simple selection for pesticide resistance. Even with proper use, pesticide resistance can be expected to emerge if use is prolonged. IPM prolongs the period that a pesticide can be used effectively, because of the complementation of various control techniques. When the bionomics of a pest are understood, it becomes apparent that various natural mechanisms impinge to control its population. Some of those mechanisms can be introduced and fostered in an agricultural setting. Biological mechanisms, such as the pathogens, parasites and predators of the pest, can substitute for synthetic chemical pesticides and can be used in conjunction with them. Thus the diversity of selection pressure on the pest species is enhanced and that impedes the evolution of resistance.

The course lectures conveyed the necessity of monitoring pest populations in IPM. Scouting a field and collecting information about a pest contributes to knowledge of the biological attributes of pest outbreaks, the human activities that foster outbreaks and the ecological factors which heighten or dampen pest populations. To illustrate those principles, life cycles of pest organisms in crops were examined. For example, most pests have particular life stages that are especially destructive to a crop. Sometimes those stages are the ones most amenable for directing control measures, but sometimes other stages are most vulnerable. For root-knot nematodes, the migratory phase and eggs of *Meloidogyne spp*. are vulnerable to control measures, but the most damaging stages are immune because they are sequestered in the roots.

Basic study of pest life cycles also allows monitoring of environmental conditions which are conducive to the development of high pest populations. That knowledge enables the farmer or extension technician to form a prognosis of required levels and timing of control and appropriate methods of pest management. The monitoring allows the grower or pest management scout to determine if the pest is likely to breach economic thresholds beyond which investment in pest control would be offset by returns at harvest.

A major focus of any discussion of biological control is the array of species which parasitize, infect, or prey upon the target pest. Such organisms have been introduced or fostered with success in many cases (Cooke & Baker, 1983;

Campbell, 1989). Life cycle studies of the pest and its control agents allow understanding of the life stages of a pest that are most susceptible to control. If biological control is to be used, the agent must be applied when the pest is present generally or present in the life stages which the bio-control agent can use. Life cycle study and monitoring of naturally occurring bio-control agents also contribute to the farmer's knowledge of when pathogens, parasites and predators of a pest are common in a field. Without this knowledge, applications of synthetic chemical pesticides may inadvertently destroy these beneficial organisms. Monitoring beneficial organisms may also allow growers to implement techniques to augment their populations and hence effectiveness. The lectures established the rationale for appropriate data acquisition and examined the ways in which the life cycle of pest organisms and their agents of biological control can be monitored most effectively in a pest management regimen.

Plant parasitic nematodes are well suited to control through IPM strategies and methods of biological control. The field and laboratory sections of the course took the methods discussed in the classroom and put them to practical use in the field. A day was spent in a local farmer's tomato field, exercising the practical methods of monitoring nematode infestation.

Figure 13.4 Instructor and student scoring nematode disease infestation

The course participants worked on a system of randomly sampling a field to determine the areas of greatest prevalence of root-knot nematodes. A method of rating plants for severity of infestation was demonstrated (Figure 13.5). The index for scoring was as follows:

- 0 = no galling (nematode free)
- 1 = 1-5 galls
- 2 = 6-25 galls
- 3 = 26-100 galls
- 4 = 100 galls singly
- 5 = many galls fused into compound galls
- 6 = compound galls with slight reduction of secondary roots
- 7 = secondary root growth strongly retarded
- 8 = secondary roots distorted and decayed (Daulton & Nusbaum 1961)

Figure 13.5 Map of tomato field demonstrating severity of nematode disease

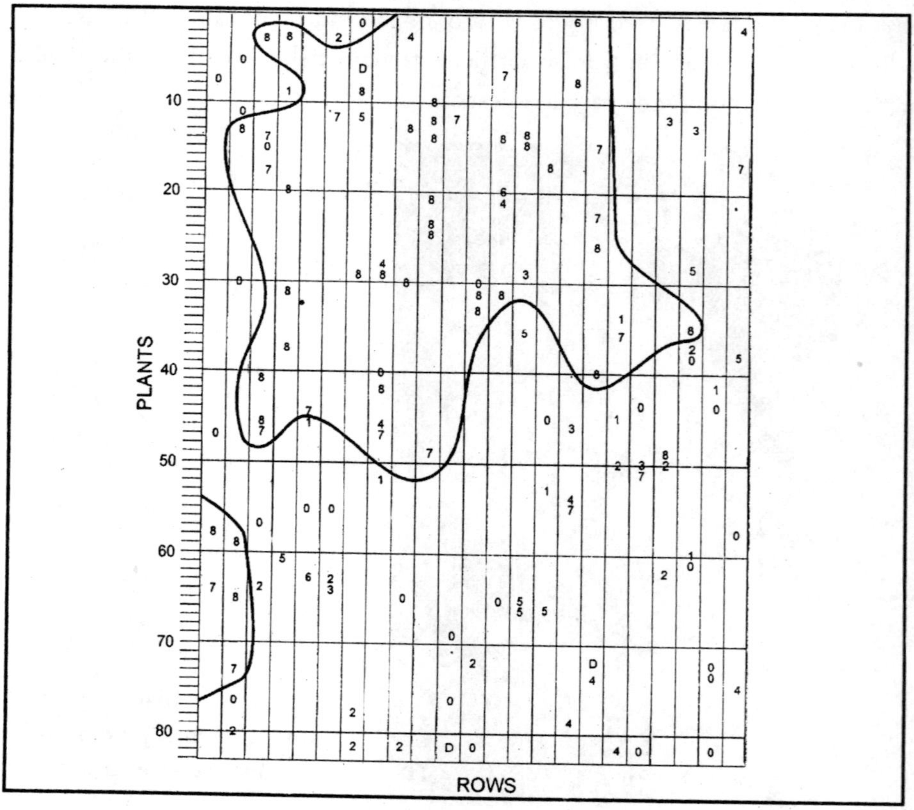

The two exercises allowed the students to produce a report demonstrating the extent of disease infestation and indicating the areas of the field which would most benefit from control tactics, or which should be avoided for future plantings of tomatoes. In the laboratory, the students learned how to extract nematodes from the soil and to determine which species of plant parasitic nematode are important. Students examined samples taken from the plots used in the research portion of the project. This field exercise practically demonstrated how nematological research can be conducted, and examined examples of the biological control of nematode pests that are amenable to resource-poor farmers in the region.

CONCLUSION

Both components of the project funded by the ECEP can be said to have been a great success. Students were requested to complete a course evaluation at the end of the week. They conveyed that much useful information was presented during the week. They said the course gave them valuable and practical experience for their research or extension activities that would enable them to more effectively manage the problem with plant parasitic nematodes in the country. The intensive nature of this double-pronged research and course effort will contribute to long-term food security and poverty alleviation for the future of Namibia.

REFERENCES

Agrios, G.N. 1997: *Plant Pathology*. 4th edition. San Diego: Academic Press.

Bartschi, H. 1982: Influence of Pesticides on Endomycorrhizal Infections. *Colloq. INRA.* 13, 259-266.

Behrens, R.H. and Dukes, D.C.D. 1986: Fatal Methyl Bromide Poisoning. *British Journal of Industrial Medicine.* 43, 561-562.

Bird, G.W. 1987: Role of Nematology in Integrated Pest Management. In J.A.Veech and D.W. Dickson (eds.), *Vistas on Nematolgy.* Hyattsville: Society of Nematologists.

Bull, D. 1982: *A Growing Problem: Pesticides and the Third World Poor.* Oxford: OXFAM.

Campbell, R. 1989: *Biological Control of Plant Pathogens.* U.K.: Cambridge University Press.

Cooke, R.J. and Baker, K.F. 1983: *The Nature and Practice of Biological Control of Plant Pathogens.* St. Paul: American Phytopathological Society.

Daulton, R.A. and Nusbaum, C.J. 1961: The effect of soil temperature on the survival of the root-knot nematode *Meloidogyne javanica* and *M. hapla. Nematologica* 6, 280-294.

Georghiou, G.P. and Taylor, C.E. 1977: Genetic and Biological Influence in the Evolution of Insecticide Resistence. *Journal of Economic Entomology*. 70, 319-323.

Hine, C.H. 1969: Methyl Bromide Poisoning. *Journal of Occupational Medicine*, 11, 1-10.

Hurtt, M.E., Morgan, K.T. and Working, P.K. 1986: Histopathology of Acute Responses in Rats Exposed by Inhalation to Methyl Bromide. *Pharmacologist*, 28(3) 207.

Kevan, P.G., Thomas, V.G. and Belaoussoff, S. 1997: AgrECOLture: Defining the Ecology in Agriculture. *Journal of Sustainable Agriculture*, 9(2/3), 109-129.

Letz, G.A., Pond, S.M., Osterloh, J.D., Wade, R.L. and Becker, C.E. 1984: Two Fatalities after Acute Occupational Exposure to Ethylene Dibromide. *Journal of the American Medical Association*, 25(2), 2428-2431.

Luc, M., Siroka, R.A. and Bridge, J. 1990: *Plant Parasitic Nematodes in Tropical and Sub- tropical Agriculture*. CAB. U.K.

Luck, R.F., van den Bosch, R. and Garcia, R. 1977: Chemical Insect Control, a troubled strategy of pest management. *Bioscience*, 27, 606-611.

Menge, J.A. 1982: Effect of Soil Fumigants and Fungicides on Vesicular-Arbuscular Fungi. *Phytopathology*, 72(8), 1125-1132.

Noe, J.P. 1988: Theory and Practice of the Cropping Systems Approach to Reducing Nematode Problems in the Tropics. *Journal of Nematology*, 20(2), 204-213.

Ohr, H.D., Munnecke, D.E. and Bricker, J.L. 1973: The Interaction of *Armillaria mellea* and *Trichoderma spp*. as Modified by Methyl Bromide. *Phytopathology*, 63, 965-973.

Starr, J.L. and Kenerley, C.M. 1986: Nematicide Evaluation as Effected by Disease Complexes and Nematicide Effects on Non-Target Organisms. In Kenneth Hickey (ed.), *Methods For Evaluating Pesticides For Control of Plant Pathogens*. U.S.A.: APS Press.

Swift, M.J., Heal, O.W. and Anderson, J.M. 1979: *Decomposition in Terrestrial Ecosystems*. Los Angeles: University of California Press.

Torkelson, R. and Oyen, F. 1977: The Toxicity of 1,3-Dichloropropene as Determined By Repeated Exposure of Laboratory Animals. *American Industrial Hygiene Association Journal*, 38, 217-223.

Whorton, M.D., Krauss, R.M., Marshall, S. and Milby, S. 1997: Infertility in Male Pesticide Workers. *Lancet*, 2, 1259-1261.

World Bank. 1992: *Poverty Alleviation Through Sustainable Growth*. Washington, D.C.: World Bank.

Yaron, G., Janssen, G. and Maamberua, U. 1992: *Rural Development in the Okavango Region of Namibia: An Assessment of Needs, Opportunities and Constraints*. Namibia: Gamsberg MacMillan Publ. Niser and OXFAM Canada.

CHAPTER 14

ENVIRONMENTALLY FRIENDLY INSECT CONTROL FOR COFFEE PRODUCTION IN ZIMBABWE

By Dumisani Kutywayo

INTRODUCTION

Arabica coffee, *Coffea arabica* L., is an important export crop grown by both small- and large-scale farmers in Zimbabwe. Coffee ranks third after tobacco and horticulture in the agricultural sector's foreign currency earnings. Smallholder plantations range from 50 to 1,000 trees per hectare and arable area committed to coffee ranges from 0.1 hectares to 5 hectares. Most of the smallholders are organized into co-operatives in the Honde Valley, Chikukwa, Rusitu Valley, Cashel Valley and Chinyaduma areas of Manicaland province. The smallholder sector contributes two to three percent of national coffee production. Most of the coffee grown in Zimbabwe is exported (99%) to Europe and the Americas, earning the country much needed foreign currency. Crop yields average about 0.6 tonnes in the smallholder sector and 1.2 tonnes per hectare in the large-scale sector.

The economic production of coffee in Zimbabwe is influenced by the availability of water for supplemental irrigation and the ability to manage insect pests, diseases and weeds effectively. Pests and diseases have been managed mainly through the use of pesticides. Over the years, Zimbabwe's commercial coffee production has relied on pest management technologies generated at the Coffee Research Station. These technologies require a lot of capital input and are specific for each management practice. It is estimated that 45% of total input costs for coffee production is allocated to pest and disease control. The country spends a lot of foreign currency to import pesticides and these high costs are beyond the reach of the smallholder sector. Pesticide use in coffee production has resulted in detrimental side effects on the environment and non-target organisms, such as parasitoids and predators.

For these reasons, there is a need to develop a holistic approach which encompasses the whole pest complex in coffee and emphasizes the generation and adoption of technologies such as biological control, host plant resistance and the integration of other environmentally friendly control measures aimed at reducing pesticide dependency. This paper describes current insect pest control measures for coffee in Zimbabwe and measures being undertaken to develop more environmentally friendly techniques.

Insect Pests of Coffee

Le Pelley (1968) listed many insect pests of coffee. In Zimbabwe, about 30 insects have been recorded. These pests can be divided into major and minor pests.

Major insect pests

The four major insect pests in Zimbabwe coffee production are coffee leaf miner (CLM), *Leucoptera meyricki* (Ghesquiere); antestia bug, *Antestiopsis orbitalis bechuana* (Kirkaldy); white stern borer, *Monochamus leuconotus* (Pascoe) and giant looper, *Ascotis selenaria reciprocaria* (Walker).

- **Coffee leaf miner (*L. meyricki*):** The leaf miner is an important pest in both large-scale and smallholder farming systems. CLM damage can lead to severe defoliation and die-back. It is believed that the leaf miner changed from minor pest status to major as a result of increased use of mulching and copper-based fungicides (Crowe, 1964).
- **Antestia bug (*Antestiopsis orbitalis bechuana*):** Antestia bug is a sucking pest which causes blackening of the flower buds, fall of immature berries, rotting and zebra-striping of beans. The injured berries are of low appeal and often have minimal market value in competitive markets (Le Pelley, 1942). Because of the direct nature of antestia damage and its subsequent effect on coffee quality, antestia has been assigned very low economic threshold levels of one to two adults or nymphs per tree.
- **Giant looper (*Ascotis selenaria reciprocaria*):** The giant looper caterpillar has never been recorded in the smallholder sector, but there are frequent outbreaks in the large-scale sector. This is because its natural enemies are decimated due to frequent spraying of parathion. Wheatley (1963, 1964) reported that giant looper occurred in estates where frequent organophospate

insecticide sprays were applied. The first outbreaks of the giant looper in Zimbabwe were at an estate where parathion was used frequently.

- **White borer (*Monochamus leuconotus*):** White stem borer used to be regarded as a minor pest and a symptom of poor management (Clowes & Hill, 1981). The neglect of coffee plantations and the 1992 drought led to an increase in white borer infestation levels. The fall of the coffee industry in Natal, South Africa has been attributed to white stem borer infestation and the outbreak of coffee leaf rust caused by *Hemileia vastatrix* (Fuller, 1901).

Minor Pests of Coffee

A number of other insect pests occur on coffee in Zimbabwe. They include the green scales (*Coccus spp.*), snout beetles (*Systates spp.*), jelly grub (*Niphadolepis alianta*), stinging caterpillars (*Latoia vivida, Astatoplebia nubilata, Neomocena convergens*), yellow-headed borer (*Dirphya nigricornis*), black borer (*Apate monacha*), semi-looper (*Chrysodeixis chalcites*), dusty surface beetle, (*Gonocephalum simplex*) berry moth (*Prophantis smaragdina*) and the coffee root mealy bug (*Planococcus fungicola*).

Chemical Control for Pest Management

Chemical control is the most common method used to manage the major insect pests (Clowes & Hill, 1981). The large-scale sector follows a routine, monthly fungicide spray programme to suppress coffee leaf rust. The decision to spray against insect pests depends on the numbers of the pest obtained at scouting. Scouting is done on 40 trees per unit of land, twice a week in summer and once a week in winter. A standard scouting form has been developed, where the cumulative pest numbers are recorded. Most of the spraying is done with hand lances and motorized mist blowers. The practice of mixing insecticides with the routine fungicide sprays has led to increased numbers of giant looper in the large-scale sector. In the smallholder sector, routine fungicide sprays are done with reduced insecticide sprays, because there is a shortage of labour for scouting and inadequate capital to buy insecticides. One interesting feature in the smallholder sector is the total absence of the leaf eaters (giant loopers, semi-loopers, jelly grub), suggesting that they are under natural biological control.

The economic threshold levels used to determine frequency of sprays against antestia bug, coffee leaf miner and other caterpillars are tabulated in Table 14.1.

These figures differ according to the age of the crop. They have not been verified experimentally in Zimbabwe, but are based on figures from East Africa, taking into account the differences in climatic conditions (Blair & Osborne, 1985).

Table 14.1 Economic Threshold Levels For Coffee Insect Pests

AGE OF COFFEE	THRESHOLD LEVELS		
(years in land)	COFFEE LEAF MINER MOTHS	ANTESTIA NYMPHS/ADULTS	CATERPILLARS
1st	2 - 5	1 - 2	1
2nd	5 - 10	1 - 2	2 - 5
3rd	10 - 20	1 - 2	5 - 10
4th	20 - 40	1 - 2	10 - 20

Most of the insecticides registered for leaf miner control are organophosphates. Parathion is the most commonly used because it is cheaper than monocrotophos, fenitrothion, trichlorfon and fention. The synthetic pyrethroids fenvalerate and deltamethrin are also recommended, but a maximum of only two sprays per year are allowed, in order to conserve the natural enemies of the leaf miner. Granular formulations of the insecticides disulfoton, aldicarb and carbofuran are also used for leaf miner control. The granules are applied in October, so that they will be taken up by the plant before the leaf miner's peak activity time in late November/early December. If the CLM population reaches the economic threshold levels after the application of a granular insecticide, but before it is effective in the plant, contact sprays with organophosphates are used to control the pest.

For antestia bug control, an average of one or two adults or nymphs per tree warrants spraying with parathion, monocrotophos, fenitrothion, fenthion, chlorpyrifos, chlorpyrifos-methyl or methomyl. Because of the direct nature of antestia damage, farmers spray regularly just to keep it under control.

White stem borer control involves application of an insecticidal deposit on coffee stems just before the beginning of the rains. The insecticide is brushed onto the stem or sprayed from a knapsack to a height of one metre above ground level. At the onset of the rains, the emerging adults, ovipositing females, eggs and subsequent larvae are exposed to the insecticidal deposit.

Only two insecticides are currently registered for the control of giant looper. These are the carbamate (methomyl) and the organochlorine (endosulphan).

Control of minor pests such as stinging caterpillars depends on their infestation levels (Table 14.1). Parathion is the only insecticide currently recommended for spraying these pests.

Reliance on chemicals to control coffee pests has given rise to a number of problems:

- worker poisoning
- environmental pollution
- destruction of beneficial non-target organisms (parasitoids and predators)

This last problem leads to secondary outbreaks of pests that are normally under natural control. For example, outbreaks of a leaf skeletoniser (*Odites artigena*) and the giant looper (*Ascotis selenaria reciprocaria*) have been attributed to the injudicious use of insecticides (Clowes & Hill, 1981).

There have been cases of pests becoming tolerant to insecticides, resulting in the use of double or triple application rates. For example, antestia bug populations collected from plantations where parathion was frequently sprayed showed greater tolerance to parathion than those collected from less frequently sprayed fields (Bisset & Van Rooyen, 1995). In 1995, there was a serious problem of antestia bug populations failing to respond to the recommended rate of parathion. Despite the disadvantages of chemicals, they still remain the only viable option for economic coffee production.

Integrated Pest Management

Because of the realization that pests can develop tolerance to insecticides, integrated pest management is now being adopted. Farmers are beginning to pay attention to both familiar and innovative non-chemical means of control and their integration into effective pest management strategies.

Cultural control

Cultural control refers to the use of good agronomic practices to maintain pest population levels below damaging levels. For example, irrigation and adequate fertilizers are recommended as a control measure against borers. For antestia bug control, the removal of shade trees and increasing inter-row spacing are recommended. Pruning is recommended for antestia control in East Africa, but is not practised in Zimbabwe, because of the risk of spreading *Fusarium* bark disease, which is a major problem in the main coffee-growing areas in the Eastern Highlands.

The root mealy bug (*Planococcus fungicola*) used to be controlled by applying dieldrin to the bared roots. Since dieldrin is no longer available, some work was done to find alternative control measures. Improving drainage by ripping between the rows and applying irrigation together with phosphate fertilizers proved to be much better than applying diazinon an alternative to dieldrin. It has also been found that planting coffee straight into untilled land previously cropped to coffee increased the infestation of snout beetle more than planting into mulched, tilled land.

Physical control

A number of physical control measures are recommended, such as hand picking and destroying stinging caterpillars and loopers. For the control of white stem borer, infested trees are uprooted and burnt. Wilted primary branches infested by the yellow-headed borer are cut off and burnt. A springy wire is widely used to kill black borer adults.

Host plant resistance

Artificial infestation of coffee leaf miner on SL28, Cattura, K7 and SL9 varieties showed that SL9 and Cattura were more tolerant than the others. More work is still being done. There is a great potential for combining host plant resistance with biological control in smallholder situations, as both techniques are environmentally friendly.

Botanical insecticides

Some laboratory work has been done to evaluate the efficacy of plant extracts such as neem oil (extract from *Azadirachta indica*), jatropha oil (*Jatropha curcas*) and the aqueous extract of *Lippia javanica*. Antestia bug nymphs and second instar larvae of giant looper were susceptible to neem and jatropha oils. *L. javanica* appeared to repel antestia bugs. More work is still being done. Plant extracts could help to replace expensive organic pesticides, particularly for the resource-poor smallholder farmers.

Growth regulators

The search for replacements for organophosphate insecticides is in progress. Insect growth regulators, such as the acyl ureas, are being evaluated for their efficacy in controlling the coffee leaf miner and other pests. Because of their specific action and minimal effect on non-target organisms, insect growth regulators are suitable for inclusion in integrated pest management programmes.

Biological Control

However, the most significant component of any integrated pest management strategy will be the introduction and management of a biological control programme. Biological control holds promise for the control of coffee insect pests because coffee is a perennial tree, providing a stable habitat for a succession of generations. The CLM has about eight generations per year in Zimbabwe (Plant Protection Research Institute, 1975) while antestia bug has three to four. It is in such habitats that biological control has occurred (Hill, 1975). The indirect pests CLM and giant looper are more appropriate targets for biological control than the antestia bug, which is a direct pest (Coppel & Mertins, 1977). Research on biological control has involved identifying the natural enemies of various coffee pests and studying the population dynamics of the coffee leaf miner and its parasitoids. Kutywayo (1989) listed the natural enemies of coffee insect pests in Zimbabwe. The coffee leaf miner is the most heavily parasitised, followed by loopers and stinging caterpillars. Other work has involved evaluating the microbial insecticide *Bacillus thuringiensis* (B.t.)

- **Coffee leaf miner:** For the coffee leaf miner, 12 larval parasitoids were identified and the most abundant of these, in descending order of importance, are *Parahormius leucopterae (Nix.), Cirrospilus variegatus (Silv.), Chrysonotomia ritchiei (Ferr.) and Pediobus coffeicola (Ferr.)*. The overall parasitism of leaf miner larvae in unsprayed coffee was found to be as high as 90 percent compared to 57 percent in sprayed coffee at the Coffee Research Station in Chipinge (Plant Protection Research Institute, 1975, 1976). Despite the high levels of parasitism, the leaf miner still poses a problem. This could be due to the destruction of natural enemies by frequent and improperly timed application of insecticide sprays. Several approaches can be taken to establish a biological control programme for CLM.

  One approach would be to undertake a classical biological control project involving exploration of *L. meyricki* parasitoids in East Africa. This is very expensive and would not yield much as the parasitoid complex of CLM in Zimbabwe does not differ much from that in East Africa (Kutywayo, 1989). A more feasible approach would be to encourage the use of granular systemic insecticides, which helps conserve natural enemies by reducing the number of contact sprays. A third approach would be to establish an augmentation programme involving the most abundant and widely distributed parasitoid, *P. leucopterae*. A mass rearing technique for CLM was developed by Katiyar and Ferrer (1968). This technique could be used to start the mass rearing of *P. leucopterae* under greenhouse conditions so that parasitoid adults can be liberated in the field just before CLM peaks in early November. In order to establish an augmentation programme for CLM, there is a need to determine the economic threshold level and study in detail the biology of the parasitoid, *P. leucopterae*. A greater understanding of the biology of the parasitoid would help in the development of the rearing technique. Conservation and augmentation seem to be the most feasible and inexpensive ways to start biological control programmes for the management of CLM in Zimbabwe.

- **Antestia bug:** The eggs of antestia bug are heavily parasitised by *Gryon sp.* and *Trisssolcus sp.* in the field, but the pest still poses a problem.
- **Giant looper:** The giant looper is attacked by a number of parasitoids and predators which keep it under control except when there is frequent spraying with organophosphate insecticides. The predators, *Macroraphis acuta* and *Rhinocoris segmentarius*, are very important in regulating the numbers of giant loopers and other caterpillars. Evaluation of commercial formulations of B.t. (Dipel and Biobit) for their efficacy in the control of giant looper larvae is being done. It has been shown that the populations of the looper's natural enemies are greater where B.t. has been sprayed than where endosulfan or methomyl have been used. Therefore, B.t. will soon be recommended for its advantage of conserving natural enemies. With respect to the giant looper and other leaf eaters, the two predators mentioned here have the potential for being used as inoculative agents to suppress larval populations when localized outbreaks occur. The predators could be mass reared in the laboratory and liberated in the field. Mugo (1992) reported on a simple method for mass rearing the predators on their hosts in the laboratory. Therefore, inoculation of laboratory reared predators seems to be a simple way of introducing a biological control programme for the control of giant looper. Finally, a nuclear poly-hedrosis virus attacking the semi-looper (*C. chalcites*) has been identified and farmers can prepare their own virus from infected larvae collected in the field.

- **White borer:** Only one larval parasitoid (*Iphialux varipalpis*) has been identified in Zimbabwe and predation by termites has recently been observed. Work on the possible use of entomopathogenic nematodes as biological control agents has recently started. Laboratory inoculation of white borer larvae with a Zimbabwean strain of *Steirnenema sp*. has proved successful. Because the white borer is capable of killing the coffee tree, reliance on biological control is questionable. However, other agents such as the entomopathogenic fungus *Beauveria bassiana* could be tried. Schoeman (1994) reported successful infection of early ringbarking larvae of white borer by *B. bassiana*.

Future biological control research options

There is a need to carry out work on the potential to establish biological control programmes for the coffee leaf minor and giant looper in Zimbabwe. Such projects have high chances of success because coffee is a perennial crop providing a stable habitat for the successive development of many generations of the pest each year. Both CLM and giant looper are indirect pests, making them suitable candidates for biological control. For CLM, biological control seems to be possible through conservation of natural enemies by using granular pesticides. Another approach is augmentation of *Parahormius leucopterae* in the field by releasing greenhouse reared parasitoids just before the CLM population peaks in November/early December. Classical biological control of CLM is ruled out because of the high costs involved in foreign exploration and quarantine. The predators *Rhinocoris segmentarius* and *Macroraphis acuta* have potential as bioagents in the control of giant looper. These predators could be mass reared in the laboratory and inoculated each time the economic threshold level of the giant looper is exceeded. Their advantage is that they are not host specific, hence can survive on other larvae of the lepidoptera in the absence of giant looper. There is a need to conduct studies on the biology of the natural enemies of coffee, the effects of insecticide sprays on population levels of parasitoids and predators, and the rate of predation due to *R. segmentarius* and *M. acuta*. Such studies would assist the proper design and evaluation of biological control programmes. Trapping adult insect pests using pheromones is an option worth trying for giant looper, berry moth and possibly the white stem borer.

Conclusion

A number of insect pests attack coffee crops in Zimbabwe. Economic production of coffee for export relies largely on the ability to manage insect pests properly. Antestia bug, coffee leaf miner, white stem borer and the giant looper are the major pests. Control of these pests is largely chemical, but this has led to outbreaks of minor pests such as *Odites artigena*. The infestation of giant looper is related to the misuse of organophosphate insecticides. Integrated pest management involving the integration of host plant resistance with biological control and other methods is advocated because it is friendly to the environment and to non-target organisms. Work on the identification of natural enemies of insect pests, host plant resistance, botanical insecticides and evaluation of softer insecticides is in progress.

References

Bisset, J.L. and Van Rooyen, M. 1995: Antestia Control in Coffee. *Preston Coffee Estates, Coffee Growers' Newsletter*, April 1995. Harare: Coffee Growers' Association.

Blair, B.W. and Osborne, R.G. 1985: Some guidelines for pest control based on Scouting and Threshold levels. In: M. St. J. Clowes and W.J.C. Logan (eds.), *Advances in Coffee Management and Technology in Zimbabwe 1980-1985*. Harare: Coffee Growers' Association.

Clowes, M. St. J. and Hill, M.R. (eds.). 1981: *Coffee Handbook*. Harare: Coffee Growers' Association.

Coppel, H. and Mertins, J.W. 1977: *Biological Insect Pest Suppression*. New York: Springer-Verlag.

Crowe, T.J. 1964: Coffee leaf miners in Kenya H. Causes of outbreaks. *Kenya Coffee*, 29 (342) 223.

Hill, D.S. 1975: *Agricultural insect pests of the tropics and their control*. First Ed. Cambridge: Cambridge University Press.

Fuller, C. 1901: *First Report of the Government Entomologist*. Natal Department of Agriculture 1899-1900.

Katiyar, K.P. and Ferrer, F. 1968: Rearing technique, biology, and sterilization of the coffee leaf miner, *Leucoptera coffeelia* (guer.). *Isotopes and Radiation in Entomology*. Proceedings of the International Atomic Energy Agency, Vienna, Dec. 1967.

Kutywayo, D. 1989: An annotated list of parasitoids and predators of coffee insect pests in Zimbabwe. *Zimbabwe Journal of Agricultural Research*, 27, 11-20.

Le Pelley, R.H. 1942: The food and feeding habits of Antestia in Kenya. *Bulletin of Entomological Research*, 33, 71-89.

Le Pelley, R.H. 1968: *Pests of coffee.* London: Longman.

Mugo, H.M. 1992: Possibility of using natural enemies to control giant looper, *Ascotis selenaria reciprocaria* in Kenya. *Kenya Coffee*, 57, 1453-1455.

Plant Protection Research Institute. 1975: *Annual Report 1974-75.* Salisbury: Department of Research and Specialist Services.

Plant Protection Research Institute. 1976: *Annual Report 1975-76.* Salisbury: Department of Research and Specialist Services.

Schoeman, P.S. 1994: Integrated control of Arabica coffee pests (part two): An overview of measures to control the white coffee stemborer. *Subtropica*, 12, 11-20.

Wheatley, P.E. 1963: The Giant Looper, *Ascotis selenaria reciprocaria* (Walk.) *East African Agricultural and Forestry Journal*, 29 (2), 143-146.

Wheatley, P.E. 1964: The Giant Looper. *Kenya Coffee*, 29, 141-154.

CHAPTER 15

NATURAL PESTICIDES IN TANZANIA

by Abel M. Mbwile

INTRODUCTION

Any organism harmful to people, animals or crops is a pest or pathogen. It has been estimated that pests and pathogens may account for a nearly 50% reduction in global harvest yield of some crops. A pesticide is anything that controls, destroys or damages harmful organisms. Many of the numerous chemicals used as pesticides enter the environment and harm non-pest species. Wide-spectrum pesticides stop the food chain by killing many different organisms and reducing biological diversity. When chemical resistance emerges, there may be a rapid proliferation of a particular pest. Chemical pesticides affect human beings and can accumulate in the human body, with residual effects that cause diseases and suffering at a later stage. Chemical pesticides are also very expensive for many small-scale farmers. Natural crop protection is the alternative. It maintains the food chain and supports biological diversity of predators and pests.

Our experience has taught us that not all modern knowledge or practice is beneficial. Developing countries can take cautionary lessons about modern farming from Europe and America. These areas now face massive problems of air and soil pollution, water contamination and soil compaction as a result of heavy application of industrial fertilizers, chemicals and machinery. Farmers should start moving back to organic farming to minimize or completely abandon chemical products in their farms. This paper shows some natural pest control methods farmers use in Mbozi District, Tanzania. Most of the analysis is set out in a tabular format showing plants used, pests controlled, methods, dosages and appropriate application sites.

NATURAL CROP PROTECTION

In traditional agriculture, farmers used locally available material to control the majority of pests. The knowledge of plant protection grew over many generations and was modified to fit local problems. The protective planting materials were selected from their gardens and were therefore adapted to the local agroecosystems. Although farmers had limited knowledge about invisible pests such as viruses and bacteria, they were quite knowledgeable about more visible pests such as insects, weeds and birds and they built their control methods around these. Plant extracts and ash were often used at various levels of production to control pests. Gradually, these practices have been discarded and forgotten; most of the useful traditional knowledge has been lost. But local materials and practices can be revived to contribute to an appropriate pest management strategy, as well as minimising the risks for farmer, consumer and the environment.

Prevention is better than cure. The emphasis should be on proper cultural practices to encourage healthy plants. A healthy soil produces healthy plants and healthy plants resist attack from pests and diseases. A well-fed soil will supply a balanced diet, and plants given a good diet are just like people – they will be much more resistant to pests and disease. The use of compost and mulching contribute to soil health and fertility. Organic manures improve the structure of the soil and this encourages plant growth.

Natural crop protection can also be achieved through cultural practices. Changing the micro-climate helps to break the pest cycle. Reintroduction of plant diversity, crop rotations, mixed cropping, intercropping, agroforestry and other cultural practices help to reduce pest outbreaks and serious damage to crops. Using pest-resistant seed and collecting seed that appears to be pest-resistant will also help. Farm cleanliness will expose pests' hideouts and helps to destroy them by interrupting their life cycle.

The encouragement of pest predators and their host plants will also serve to control pests. Removal of the natural vegetation on farms and around gardens has greatly disturbed the pest/predator balance. The essential food supplies of birds, animals and insects have been reduced. Very often the natural vegetation has been so depleted by deforestation and overgrazing that natural predators have no vantage points and the pests can operate freely. Reintroduction of plant diversity will support a greater balance in pest/predator populations.

If a plant is unhealthy, the problem must be identified before an effective cure can be found. It is important to understand the biological interdependence and relationships that exist in the field. Both destructive and beneficial organisms must be identified. The biology of pests and beneficials, their body structure, life cycle, behaviour, breeding and feeding patterns are all important to understanding how to promote crop health. This information can help determine the best time to attempt to destroy pests or protect against them. The relationship between crop stages and pest outbreaks is thus important.

A variety of methods can be used to repel and kill pests. Barriers can be used to protect plants. For example, a square of carpet laid on the soil around the base of a cabbage plant will fool the cabbage root fly. A bottomless container from a plastic oil bottle will keep slugs and snails away from young plants. Traps can be used to catch rodents and small mammals. Trap crops provide an alternative host which can keep pests away from the main agricultural crop. For example, milkweed attracts aphids which may otherwise go to a crop such as cowpeas. Larger pests can be picked off the plants by hand. The introduction or encouragement of predators such as snakes, lizard, chameleons, wasps, spiders and birds can be encouraged. A lamp can be left on at night to discourage night feeders.

Botanical Controls

Physical control can be supplemented by botanical control in the form of natural pesticides sprays and powders. Natural pesticides break down quickly without leaving any harmful by-products and they are locally available. Nature offers us a profusion of plants for controlling pests Table 15.1 offers a list of some natural botanical controls which farmers are using in Mbozi District, Tanzania. The table shows the plants used, pests controlled, methods and the appropriate dosages and application sites.

Table 15.1 Some natural crop protection methods to destroy pests

Plant/ Scientific name	To Kill or Repel	Preparation and Application
Fish bean (*Tephrosia vogelii*)	Grain bore weevils, ants	Dry leaves under the sun. Grind into powder. Mix 120 gm of the dry powder into 90 kgs bag of grain
Fish bean (*Tephrosia vogelii*)	Tick, lice	Crush fresh leaves and soak in water for 24 hours in a ratio of 1 part of leaves to 5 parts of water. Then wash the animal with the solution (filtered or not filtered). 20 litres of solution can wash 6 cows
Fish bean (*Tephrosia vogelii*)	Aphids, cutworms, cockroach, louse, grasshopper	Pound fresh leaves of the plant and mix with 5 times quantity of water. Boil for 30 minutes. Add one teaspoon of washing soap for each litre of water so that the medicine sticks on the leaves. Then sieve or filter into a sprayer ready for spraying in field.

Garlic (*Allium Sativum*)	Chicken fungus, aphids, caterpillars, cabbage worms	1. Grind 100 gm and mix with one litre of cold water, add one teaspoon washing soap, filter, ready for spraying fungal infested plants 2. For repelling pests grow along the edges of the plant bed
Marigolds (*Tagets minuta*)	Nematodes, cutworms, caterpillars, flies, ants	Pick leaves at flowering stage, place it in a container and cover with water. Let stand in water for 5 - 10 days stirring occasionally. After five to seven days remove the large decomposed remains and use as mulch on vegetables. Then strain or filter the rest, dilute with soapy water and spray on infected plant
Comfrey (*Symphytum peregrium*)	Aphids	Grind and soak 100 gm/litre of water for 12 hours. Filter and spray on the infected plants.
Fingermillet	Weevils, borers	Mix the bran of fingermillet with grain after threshing and store
Farm Yard Manure (FYM)	Weevils, borers stophillus	Take dried FYM and burn. After cooling, take 0.5 litre of the ash and spread under the gunny bag. Put in your grains and after every 60 litres of grains put another 0.5 litre of dung ash.
Pyrethrum/ (*Chrysanthemuan Cinerariae folium*)	Aphids, white fly, spider, mite mealy bug, and soft scale rose slug	Made from pyrethrum flower liquid or powder, safe to mammals but harmful to bee and some lady birds.
Wood ash	Cutworms	If you put wood ash on the crop, the alkaline nature of ash irritates cutworms.
Chillies or hot pepper	Black ant, white ant, grain borer, larva in cabbage, rice weevil and other virus	1. Crush and then mix with water, leave for 24 hours, filter and add about 9 litres of soapy water solution and then spray. 2. For storage use 50 gm of hot pepper powder mix with 90 kg of maize for storage of grains.
Lantana camara	Potato moth during storage of potatoes	Leaves and branches contain the active ingredient. Lay potatoes on fresh lantana branches during storage.
Neem tree	Fungi and nematodes	Leaves, fruits and roots have insecticidal repellent antifeedant and pest-growth inhibiting qualities.
Mkweme	Rashes, aphids	Plant produces seed very rich in oil. Leaves are crushed and put in 10 litre of water for 12 hours, sieved and then the animal or plants are sprayed with solution.
Cock plant *Kogoligo*	Maize stalkborer	Crush roots/leaves of the plant, soak in water for 2 hours, then spray in the affected maize plants.
Nyongwe	Aphids bean fly, American ball-corm	Crush one half to one kg. of tuber. Put in 20 litres of water for 12 hours. Sieve, add teaspoon of soap then spray over your crops
Sunflower (*Helianthus annuus*)	Stalkborers, Caterpillars	Crush fresh leaves, soak in water for 12 hours. Sieve and apply in the affected plants
Ndindindi	Maize stalk borer	Crush the leaves and put in 10 litres of water. After 12 hours squeeze over the maize funnel

Tobacco (*Nicotiana tabacum*)	Leaf miners, lice, aphids, larva, moth, grainweevils, grasshoppers, borers, cutworms, caterpillars, cabbage worms	Boil the mixture of tobacco leaves for half an hour then apply the tobacco tea on the affected plants. Apply only if other natural plants do not prove effective. Tobacco tea is poisonous to human and animals. Keep away from children. Add a piece of washing soap to promote sticking to the plants.
Cypress (*Supressus Lusitanice*)	Sweet potatoes weevils, storage pests.	For storage pests cut fresh leaves/branches of cypress and put under the gunny bag. After filling, put 1 litre leaves/ branches of cypress over the bag. The strong smell will repel pests for a long time.

Potential Problems

If the use of such local botanicals is to be adopted by most farmers, there is a danger of massive genetic erosion as plants are harvested from the wild to be used as pesticides. It is important, therefore, that in advocating their use, we should also collect, protect and propagate these plant materials on an adequate scale. The culture of botanicals will have to become part of any mixed farming strategy. It may also be necessary to accept that in some cases of pest outbreak (army worms or locusts), the use of botanicals may not be adequate. It is not easy to prepare botanicals in sufficient quantity to meet such crises, and in that case we may have to resort to commercial pesticides. Even though our major thrust is the small-scale farmer, we should not overlook the fact that farmers are also ambitious, and usually like to expand their farming activities. We must look into ways to ensure there are enough botanicals to cope with increasing farm sizes.

Conclusion

The potential of natural pest control can only be fully explored when traditional knowledge is valued and integrated in current practices. Documentation of current natural pest-control practices and the collection of indigenous knowledge are important. If natural pest control is to be promoted, researchers have to learn from the farmers. More scientific investigation could do much to improve the applicability and effectiveness of botanicals. Research on their active ingredients, on application preparations and rates, and on the environmental impact of botanicals are all required. In reviving and promoting natural crop protection, it is imperative to establish the side effects of these botanical both to the farmer and to beneficial predators.

The use of natural pest control requires commitment to make the system work. The farmer must be convinced and motivated, in order to manage effectively. The formation of farmer networks and working groups may contribute to success in using natural pest control, as each farmer relies on the findings and practices of others for motivation and support.

Finally, because every farmer is interested in profit and protecting against risk, it is necessary to help organic farmers market their products. Most people are aware of the health hazards caused by industrial chemicals and they are prepared to pay more for healthy foods. Helping organic farmers secure good prices for their products will promote organic farming.

Chapter 16

South African Entomopathogens as Biological Control Agents of Vegetable Pests

by Mark G. Wright, Sehaam Khan and Sean Davison

Introduction

Integrated pest management (IPM) combines various methods of managing insect pests. Reliance on a single or a few control measures is reduced. Compatible control measures are used to keep pest populations below a threshold level. The emphasis is on management of pests rather than their elimination. Biological control is an important component of IPM, using natural enemies of pests to reduce their population numbers. These natural enemies may be predatory insects, parasitic insects, parasitic nematodes and a range of pathogens that infect and kill insects. IPM is used extensively in fruit production in South Africa, but is largely unexplored for vegetables and other horticultural crops (e.g. flowers). Commercial vegetable producers rely heavily on pesticides, although many are interested in the use of IPM. Small-scale or resource-poor farmers often do not use pesticides for various reasons, and are generally enthusiastic about the possibility of using natural pest control measures.

A major problem in applying biological control in vegetable crops is their temporary nature. Unlike deciduous fruit orchards, for example, where biological control agents become established and do not require augmentation, vegetables are annual crops that do not provide a perennial substrate for beneficial organisms. Commercial production of beneficial insects is costly, reducing the appeal of this approach.

In contrast, insect pathogenic fungi may be produced relatively cheaply, lending support to their potential value in IPM. Insect pathogens are becoming increasingly important in pest management (e.g. Parker, Skinner, Brownbridge, Adamowicz & Su,1996). Entomopathogenic fungi have great potential, as they may be cheaper to produce commercially than predators and their application is relatively easy. They are used quite extensively world-wide as biological control agents, particularly for the control of thrips, whitefly, various borers (Parker *et al.* 1996) and even migratory locusts in the dry climate of the Karoo in South Africa (Muller 1997). Some are commercially available in Europe and the United States (Malais & Ravensberg 1992). Insect pathogens are generally sensitive to climatic conditions such as drought and heat, as well as exposure to ultraviolet radiation. As dry, hot conditions and high UV exposure are characteristic of South African conditions, it would be beneficial to discover indigenous pathogens that may be adapted to these conditions. The aims of this project were:

- to determine which insect pathogens are present in cultivated and wild ecosystems in the Western Cape, South Africa
- to evaluate the pathogenicity of these isolates under laboratory and glasshouse conditions
- to evaluate the efficacy of selected isolates under field conditions.

SURVEY OF INDIGENOUS PATHOGENS

Vegetable pests were collected in various vegetable growing areas (Caledon, George, Robertson, Paarl, Stellenbosch), as well as in the wild (Anysberg Nature Reserve, Grootswartberge, De Hoop Nature Reserve, Jonkershoek Valley). The following pests were included:

- *Thrips tabaci* (onion thrips, Thysanoptera, from onions)
- *Frankliniella occidentalis* (western flower thrips, Thysanoptera, from tomatoes and flowers)
- *Bemisia tabaci* (whitefly, Hemiptera, from tomatoes and cucurbits)
- *Brevicoryne brassicae* (grey cabbage aphid, Hemiptera, from crucifers)
- *Myzuspersicae* (green peach aphid, Hemiptera, from tomatoes)
- *Helicoverpa armigera* (bollworm, Lepidoptera, from tomatoes)
- *Plutella xylostella* (diamond back moth, Lepidoptera, from crucifers)
- *Pieris brassicae* (cabbage white, Lepidoptera, from crucifers).

Indigenous Proteaceae, an element of the Cape flora, bear infructescences that host various insects, and these provide a valuable environment for insect pathogenic fungi. Insect borers collected from these were investigated for pathogens. As these insects occur in hot, dry areas, they may be valuable sources of hardy insect pathogens. Insect taxa included were:

- *Sphenoptera* spp. (flat-headed borers, Coleoptera)
- *Tinea* sp. (Lepidoptera)
- *Argyroploce* sp. (Lepidoptera).

Field-collected insects were returned to the laboratory, where any showing indications of being infested by pathogens were investigated further. The insects were surface sterilized in sodium hypochlorite (3.5 %) for 30 seconds, and rinsed in distilled water. They were then placed on corn-meal agar (CMA) in sterile petri dishes. To avoid contamination by saprophytic fungi that may have been present, insects were initially placed onto the lids of the dishes (the fungi we seek have spores that are expelled forcefully). Any pathogenic fungi that grew onto the CMA were re-isolated to obtain pure cultures for identification. Pure cultures were submitted to the ARC-Plant Protection Research Institutes' National Collection of Fungi for identification, and any insect pathogenic fungi were deposited in their voucher collection.

The survey yielded pathogens from thrips, whitefly, various Lepidoptera and Coleoptera, and other insects that were collected opportunistically (Table 16.1). A number of these isolates remain unidentified as they are from relatively unknown groups of fungi, or groups for which there are not mycologists in South Africa who can provide identifications.

Table 16.1 Pathogens collected from insects during the survey of the Western Cape.

Isolate	Taxon	Host	Locality
G1	*Verticillium lecanii*	*Thrips tabaci*	Elsenburg
G4 *C*	*Ordyceps militaris*	*Lepidoptera*	Stellenbosch
G6	*Fusarium sp.*	** Sphenoptera sp.*	Anysberg
G7	*Beauvaria bassiana*	*Sphenoptera sp.*	Anysberg
G8	*Beauvaria bassiana*	*Sphenoptera sp.*	Anysberg
SK1	*Beauvaria bassiana*	Coleoptera	Spreeufontein
SK2	*Beauvaria bassiana*	Coleptera	Spreeufontein
SK3	*Gliocladium sp.*	* Coleptera	Anysberg
SK4	*Verticillum lecanii*	*Bemisia tabaci*	Paarl
SK5	Indet.	Coleoptera	Anysberg
SK6	Indet.	Coleoptera	Jonkershoek
SK7	Indet.	Coleoptera	Jonkershoek
SK8	*Epicoccum sp. (nigrum?)**	Coleoptera	Jonkershoek
SK9	Indet.	Coleoptera	Jonkershoek
SK10	*Fusarium sp.**	Coleoptera	Du Toits Kloof

SK11	*Epicoccum sp.**	Aphididae	Elsenburg
SK12	Indet.	Aphididae	Elsenburg
SK13	Indet.	*Bemisia tabaci*	Paarl
SK14	Indet.	*Bemisia tabaci*	Paarl
SK15	Indet.	*Phytoseiidae*	Paarl
SK16	*Fusarium proliferatum**	*Thrips tabaci*	Robertson
SK17	Indet.	*Thrips tabaci*	Robertson
SK18	*Fusarium sp.**	*Thrips tabaci*	Caledon
SK19	*Paecilomyces lilacinus*	*Thrips tabaci*	George
SK20	*Fusarium sp.**	*Thrips tabaci*	George
SK21	*Fusarium sp.**	*Thrips tabaci*	Caledon
SK22	*Fusarium sp.**	*Plutella xylostella*	George
SK23	*Fuarium sp.**	*Thrips tabaci*	George
SK24	*Fusarium sp.**	*Thrips tabaci*	George
SK25	*Fusarium sp.**	*Plutella xylostella*	George
IN1	*Sporothrix sp.**	*Capys alphaeus*	Jonkershoek
IN2	Indet.	*Thrips tabaci*	Elsenburg
IN3	*Penicillium sp.**	*Brevicoryne brassicae*	Paarl
IN4	*Beauvaria bassiana*	*Bostra conspicualis*	Paarl
IN4	Indet.	*Thrips tabaci*	Elsenburg
IN6	*Beauvaria bassiana*	*Bostra conspicualis*	Stellenbosch
IN7	*Beauvaria bassiana*	*Argyroploce sp.*	De Hoop
IN8	Indet.	Scarabaeidae	De Hoop
IN9	*Beauvaria bassiana*	Lepidoptera	De Hoop
IN10	Indet.	Braconidae	De Hoop
IN11	Indet.	*Orophia ammopleura*	De Hoop
IN12	Indet.	*Conopia sp.*	Clanwilliam
IN13	*Bacillus poss. thuringiensis*	*Pieris brassicae*	Elsenburg

* = Taxa which are possibly not entomopathogenic.
Indet. = Not yet identified.

Laboratory and Field Procedures

Isolates of entomopathogenic fungi that were collected in the field were applied to captive insects to ensure that they were pathogenic. This was done by growing the fungi up on CMA, and washing spores off with distilled water. A spore suspension with 1.0 x 10^6 spores per ml was applied as a fine spray to insects placed on surface sterilized leaves in petri dishes. A water-only control was used in each case. Whitefly (*B. tabaci)* and onion thrips (*T. tabaci*) from tomato greenhouse or a captive glasshouse colony respectively, were used as assay organisms. Ten insects were enclosed in each petri dish, and each treatment consisted of at least three replicates. The petri dishes were sealed with parafilm and kept at room temperature. Mortality of insects due to infection by fungi (mycosis) was determined after ten days. All dead insects were placed on CMA again, to confirm that mortality was the result of mycosis. Pathogenicity of a *Beauvária bassiana* isolate (IN7) was determined for *P. brassicae* by applying 2 x 5 μl of a 2 x 10^6 spore per ml to large (L4) and small larvae (L2). Three larvae were included in three replicates for each size class. Control insects were treated with distilled water. Fungi from dead larvae were isolated on CMA.

Glasshouse trials were conducted on onions and broccoli infested with *T. tabaci* or *B. brassicae*. Spore suspensions were applied as a fine spray. On one occasion groundnut oil was added to the solution in an attempt to ensure that spores did not dehydrate. Other treatments were done with pure spore suspensions. Control plants were treated with distilled water. Efficacy was determined by evaluating insect survival on plants and dead insects were plated onto CMA to confirm that fungi were responsible for the mortality.

Spore suspensions were also applied to whitefly colonies on cucumbers in greenhouses. Treated leaves (with colonies of juvenile whitefly) were enclosed with fine gauze following the application. Controls were treated with distilled water. Dead insects were plated onto CMA to confirm the presence of pathogenic fungi.

In vitro bioassays were conducted to determine optimal doses of various isolates, and how long the fungi take to kill the insects. Spore suspensions of 0.6 x 10^6, 0.8 x10^6, 1.0 X10^6, 1.2 x 10^6 and 1.4 x 10^6 spores per ml were applied to two replicates of *T. tabaci* ($n = 10$) adults and juveniles in petri dishes. An untreated control (distilled water only) was included. The insects were collected from a captive colony. Insect mortality was assessed daily. All insects that died were plated onto CMA to confirm the cause of mortality. Isolates used for bioassays were *Verticillium lecanii* (SK4, GI, G16), *Cordyceps militaris* (G4) and *Beauvaria bassiana* (G8). This experiment was repeated for *M. persicae*, using isolates SK4 and IN4 and an untreated control.

In the field two isolates, *V. lecanii* (SK4) and *B. bassiana* (IN4) were applied for the control of *F. occidentalis* in a 600 m^2 greenhouse with cut-flowers. This provided the opportunity to evaluate these isolates on a non-food crop under large-scale commercial conditions. Fungi were mass-produced by first growing cultures on CMA and transferring these to double sterilized millet seed. The seed provides a carbohydrate substrate for the fungi. After 10 to 14 days, spores were washed out of the seeds with sterilized distilled water and filtered through muslin cloth. For the first trial conducted, only SK4 was applied, at a concentration of 1 X 10^6 spores per ml. This was applied to plants that were three weeks old. Thrips were collected from the greenhouse prior to application and they were placed on CMA to determine whether *V. lecanii* may have been present before spraying. Three weeks after application, we collected thrips again to determine whether SK4 had become established in the greenhouse. A second trial was conducted with a mixture of SK4 and IN4 (both 1 X 10^6 spores per ml). These fungi were applied as a soil application with planting, to establish whether this would ensure effective establishment of the fungi. This trial was still in progress at the time of compiling this report.

Results

Thrips treated with *V. lecanii* experienced up to 35% mortality and whitefly up 43%. These mortality levels were consistently significantly greater than control mortality ($P < 0.05$). *Cordyceps militaris* was ineffectual in these initial trials. All insects that had died in controls were found to be infected by the pathogen with which they had been treated. Mortality of the *P. brassicae* that were treated with *B. bassiana* 33%, and presence of the pathogen was confirmed by re-isolation. Younger larvae (L2) appeared to have been impervious to this treatment, with no mortality owing to *B. bassiana*.

Glasshouse trials gave variable results. The use of groundnut oil to prevent desiccation of spores resulted in phytotoxic responses in broccoli and the oil appeared to kill thrips more effectively by smothering them, than the fungi applied. However, treatments both with and without groundnut oil did result in infection of insects on the plants. Insect survival was, however, high. *Verticillium lecanii* appeared to establish well in the glasshouse environment, and was re-isolated from the thrips population again, two months after the trial. Insect populations were never reduced dramatically by the effects of the fungi though. The greenhouse trials gave more promising results, with whitefly populations being reduced by up to 65% by the fungi applied. Re-isolation from dead insects confirmed that the fungi that were applied had killed them.

Results of bioassays on thrips indicated that isolate SK4 (1.0 x 10^6) gave 50% juvenile insect mortality after four days. Other concentrations gave far lower mortality and at a slower rate. Results for adult insects treated with SK4 were similar, with 12 x 10^6 spores per ml giving 66% mortality after 10 days. The lower concentration of 0.6 x 10^6 gave 55% mortality after only four days. Isolate G8 only provided substantial juvenile mortality at high concentrations (1.2 x 10^6 and 1.4 x 10^6 spores per ml), and after six and seven days respectively. Adults treated with this isolate were either unaffected or all died (1.0 x 10^6 spores per ml) after seven days. Isolate G4 gave 50% juvenile mortality after 6 days (1.2 x 10^6 spores per ml) or eight days (0.8 x 10^6 spores per ml). Adult insects were less susceptible to isolate G4, although high mortality was evident after 10 days for certain concentrations. Isolate G1 was not particularly efficacious against juvenile thrips, with only one concentration giving 50% mortality after 10 days. In contrast, adults were more susceptible to this isolate, with 1.0 x 10^6 giving 100% mortality after 10 days. Control mortality was 0% percent, until one control dish became contaminated with *V. lecanii*, resulting in 12% control mortality. Isolation of fungi from dead and live insects after ten days indicated that in some cases, live insects were infected. This was observed in the case of isolates G1 (live juveniles infected), and G8 (live adults infected). Bioassay results for *M. persicae* showed high control mortality. However, the fungal treatments resulted in more rapid mortality showing that the fungi applied did have an effect.

Under field conditions thrips (*F. occidentalis*) collected prior to applying *V. lecanii* yielded no isolates of pathogenic fungi, indicating that there were none present prior to the treatment. Isolation from insects three weeks after the application showed that the fungus had become established in the insects, albeit at a low level (ca. 20% of insects yielded fungi). The application failed to prevent a huge outbreak of thrips that occurred directly after the application. This outbreak followed a period of extremely high temperatures.

Discussion

The survey of pathogens in the Western Cape revealed a number of different species of fungi. The diversity of fungi collected was not dramatic, but a number of isolates with bio-control potential were obtained. *Verticillium lecanii* and *Beauvaria bassiana* are of particular interest, as they have been shown to be valuable or to have potential for the control of pests internationally (Lindquist 1996). The discovery of a *Cordyceps militaris* isolate is also of some value, although the one collected appeared to demonstrate variable efficacy. Of great importance is that most of the isolates obtained are from arid or semi-arid areas, and they may therefore be adequately drought resistant to be of value in pest management in South Africa.

The possible *Bacillus thuringiensis* isolated from *P. brassicae* is also of great interest. Bacillus products are amongst the most effective and environment-friendly pesticides available. A local isolate with efficacy against this pest would be of great value to commercial and resource-poor farmers alike, as no control measures are currently registered for cabbage whites.

The inconsistent glasshouse results are likely to be the result of poor climate control in the glasshouses used. Relative humidity and temperature fluctuated dramatically, and this may have affected the fungi negatively, as well as creating favourable conditions for insect proliferation. However, the fungi were able to establish in the glasshouse, although at low levels. The trials conducted in greenhouses were more encouraging. These trials were possibly more effective because the greenhouses are well ventilated and relative humidity remains fairly high. It is also possible that the isolates used were more infective on whitefly than on thrips and aphids.

The *in vitro* bioassays showed that the isolates tested were able to infect thrips. SK4 and G8 gave the best results, with rapid mortality and at realistic spore concentrations. The bioassays on aphids were less conclusive (owing to high control mortality), but also showed that SK4 and IN4 were able to give good mortality in a short period. *In vitro* conditions are dramatically different to field conditions, however, and therefore only provide a guideline for possible further use.

The field application of SK4 was confounded by adverse weather conditions (high temperature), which allowed thrips to reproduce rapidly. The fungus was, however, able to become established in the trial, showing that it has potential as a biological control agent. It may be possible to combine this treatment with chemical control (using environmentally compatible pesticides), as has been the case in other studies (e.g. Lewis, Berry, Obrycki & Bing, 1996). Results from the ongoing soil application trial will indicate whether this means of application is possibly more effective.

This investigation has revealed a range of indigenous South African insect pathogens, some of which have potential value in the biological control of vegetable pests. Brownbridge (1995) suggests that the following steps should be followed to develop insect pathogens as biological control agents: selection of pathogens for screening; laboratory pathogenicity tests; select effective isolates; assess performance in controlled trials; develop mass production and application techniques; pilot test and integrate with other management approaches. The present project has reached the controlled trial phase. Mass production and other aspects have yet to be undertaken, but were beyond the scope of the project. Further development of these isolates may eventually contribute to more effective IPM and the advancement of environmentally compatible pest management in Southern and Eastern Africa.

ENDNOTES

Funding for this research was provided by ECEP and by the Professional Development Project of ARC. Identifications of fungi were done by Isabelle Ronge and W. Marassas.

REFERENCES

Lewis, L. C., Berry, E. C., Obrycki, J. J. and Bing, L. A. 1996: Aptness of insecticides (*Bacillus thuringiensis and carbofuran*) with endophytic *Beauvaria bassiana*, in suppressing larval populations of the European corn borer. *Agriculture, Ecosystems & Environment* 57: 27-34.

Lindquist, R. 1996: Microbial control of greenhouse pests using entomopathogenic fungi in the USA. In: *Insect pathogens and insect parasitic nematodes*. IOBC Bulletin 19(9): 153-156.

Muller, E. J. 1997: Update on *Metarhizium* as a biological control agent of locusts. Proceedings of the Entomological Society of Southern Africa Congress, Stellenbosch, July 1997. 86-87.

Parker, B. L., Skinner, M., Brownbridge, M., Adamowicz, A. and Su, F. C. 1996: Entomopathogenic fungi for thrips management: Recent advances. *Acta Horticulturae* 431: 521-534.

CHAPTER 17

A FARMING SYSTEMS APPROACH TO ECOLOGICAL AGRICULTURAL RESEARCH

by Clive Lightfoot and Reg P. Noble

INTRODUCTION

The Farming System Approach (FSA) refers to a form of scientific research which seeks to understand and analyse a set of interacting characteristics of operational farms. Four important variations in focus can be observed. In the early days, now some twenty years ago, the focus was on increasing yields of particular commodities. Biologists and economists worked together to identify techniques to improve commodity performance and took into consideration interactions with other components of the farming system. This enterprise or commodity focus is still popular among research station scientists seeking commodity production goals (Norman 1978; Hildebrand 1986).

In a second variant, scientists sought to use FSA to enhance the relevance of technology development for smallholders. They were concerned that smallholders were not adopting many technologies emerging from research stations. These scientists developed diagnostic procedures to enhance the participation of farmers in identifying problems. They analysed interactions between components and gave rise to analysis of crop-livestock interactions and agro-forestry. The biologists and economists working with this focus aimed to improve the production of the farming system as a whole. This focus remains, perhaps, the dominant application of the FSA today (Raintree 1984; Byerlee, Triomphe & Sebillotte 1991; Collinson 1981).

Concern for sustainability, preserving natural resources and understanding ecosystems brought a third change. This broadened scientists' understanding of the boundaries of a farming system. Resource-poor farmers rely on common property resources – forests, streams, small water-bodies and open grasslands – beyond their farm boundaries for fuel-wood, fishing and grazing land. Agroecosystems analysis sought to enhance ecological, economic and social equity dimensions of the farming system (Conway 1985). Most of the research in this area concerns the integrated management of nutrients, pests and the development of integrated farming systems (Altieri 1989; Lightfoot 1990; Kenmore 1991; Mollison. 1993; Edwards, Pullin & Gartner 1988).

The fourth variant concerns livelihood. While farming dominates the food and income security strategies of smallholders, it is not their only strategy. Poorer households send fathers, sons and daughters off to work, sometimes in far-off cities, to secure cash for the farm. In addition, non-farming pursuits are an important element of strategies for households located in the buffer zones adjacent to conservation areas The ways that households build livelihood strategies through non-farm and off-farm work and use natural resources in an ecologically sound way is the question addressed in this focus (Ellis 1998). These four versions of the FSA are depicted in Figure 17.1.

Figure 17.1: Changing focus in farming systems research

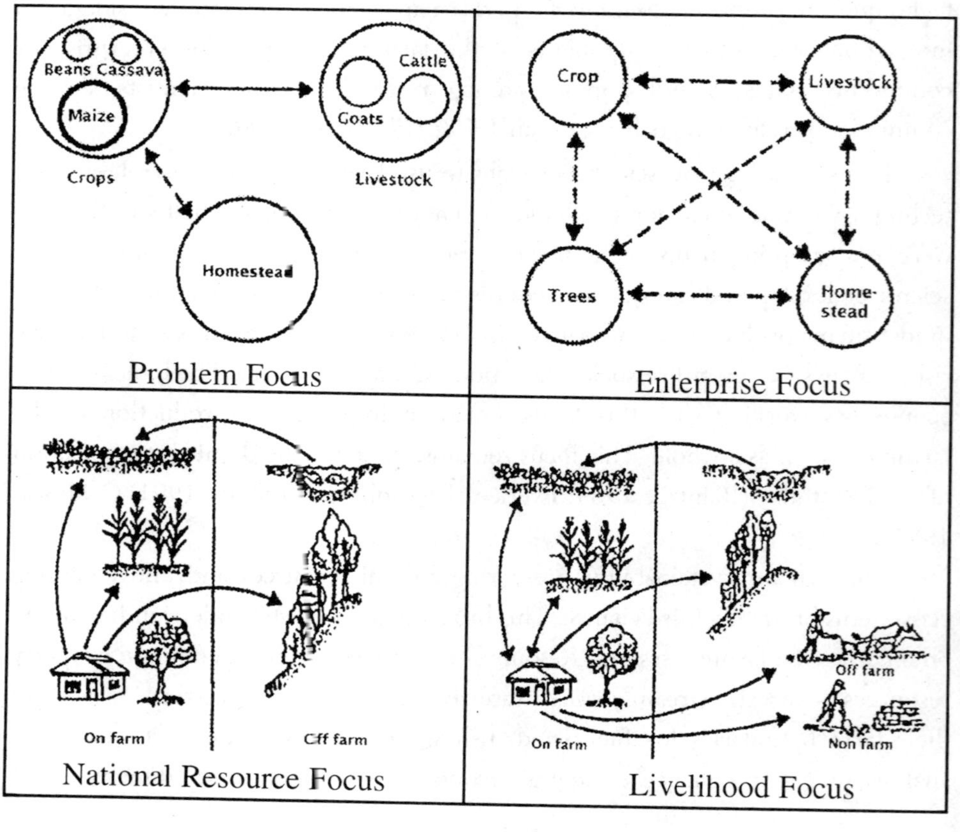

This paper presents an ecological approach to the operation and study of farming systems, drawing from agroecosystems analysis and a natural resource focus. In the following section, the basic elements of ecological management are presented. Then a framework to guide the creation and study of ecologically sound farming systems is set out. A range of sources providing evidence of the success of the approach are presented. The paper then highlights some hurdles this approach presents to scientists in national agricultural research organizations.

Analysing Agricultural Ecology

Agroecosystems analysis brought an ecological perspective to the FSA (Conway 1985; Rambo. 1984; Sajise 1986). It helped scientists understand the ecological setting of farming by breaking down the wider agricultural landscapes into smaller agroecological niches – the land types of the now-familiar agroecosystems transect, as seen in Figure 17.2. Completed transects also required the exploration of species diversity in cultured, collected or captured forms. Management of agroecological niches was further analysed through seasonal calendars and resource flow models. For the first time, farming systems scientists had tools to see the ecological dimensions – the habitats, biodiversity and food webs of farming. They were able to look beyond the commodity interest of the crop field or livestock unit. They extended system boundaries to include streams, rivers, forests, and grazing lands. The maps, transects, seasonal calendars and flow models brought a new richness to the understanding of farming systems and helped improve communication between researchers and farmers. The diagrams prepared through agroecosystems analysis were easy for farmers to understand. Indeed, as the practice of working with farmers improved and Rapid Rural Appraisals evolved into Participatory Rural Appraisals, farmers began to draw the maps and calendars themselves. Moreover, the use of local terms further facilitated communication between farmers and researchers.

Many of the ecological dimensions illuminated through agroecosystem analysis also provide information on ecological soundness or maturity. Ecological theory informs us that the following attributes are associated with ecosystem maturity:

- high functional diversity of species
- extensive recycling of by-products with positive interaction and feedback loops

Figure 17.2 Agroecosystems transect from Mampong, Ghana

	Upland	Midland	Road	Village	Lowland	Stream	Fishpond
Soil type	Gravel Stoney gravel	Gravel Loam Clay loam Clay		Gravel Loam Clay loam Clay	Clay Sandy clay	Clay Sandy clay	Clay Sandy clay
Water source	Rain	Rain		Stream, well	Rain		Stream
Crops	Cassava, Maize	Plantain, cassava maize, sugar cane yam, cocoyam		Plantain, banana	Cassava, maize, yam, cocoyam, plantain, banana		
Vegetables		Tomato		Chili pepper, eggplant, okra	Tomato, cabbage, chili pepper, bell pepper, eggplant, okra		Tomato, cabbage, chili pepper, bell pepper, eggplant, okra
Trees	Avocado, leucaena	Calliandra (nyamedual, ketawamforo)		Mango, orange	Bamboo, oil palm raffia palm, cocoa, papaya, mango, avocado		
Forages		Woodlot, grass, leaves	Grass		Grass		
Animals		Sheep, goat, chicken		Sheep, goat, chicken, duck	Sheep, goat, chicken, land snail land crab	Duck, water snail	Duck, tilapia, catfish

- high reliance on internal dynamics and ability to recover from perturbation
- high resource capacity (Odum 1969, Jorgensen 1992)

These attributes have been used as a basis for an analytical framework to compare the ecological soundness of farming systems (Dalsgaard, Lightfoot & Christensen, 1995). This matrix of farming system type against ecological attribute (Figure 17.3) provides an analytical framework to compare farming systems on their ecological soundness. One axis estimates numbers of agroecological niches, enterprises, species diversity, bio-resource recycling flows and the capacity of the natural resource base. The second axis describes four farming system types: monocropping, multiple cropping, mixed farming and integrated farming. With this analytical framework, we can imagine a gradient in ecological soundness from low to high, as we move from monocropping on a single agroecological niche, through multiple cropping and mixed farming on several niches, to intensive integrated farming on many niches with high recycling of farm wastes.

Figure 17.3: Analytical framework for ecological soundness

Farming system	monocropping	multiple cropping	mixed farming	integrated farming
Agro-ecological niche	single	few	multiple	multiple
Enterprise assemblage examples	rice or corn	rice-wheat or corn-cassava, cereal-legume	cattle, poultry, vegetables, cereals, legumes, fruit trees	livestock, poultry, fish, vegetables, cereals, legumes, aquatic plants, fruit trees, multipurpose trees
Species diversity	low	low	medium	high
Bioresource recycling	low	low	medium	high
Natural resource capacity	low	medium	high	high

Attributes of diversity, recycling and resource capacity suggest a basis for farm management strategies. Farmers can increase diversity on their farms by raising the numbers of species they cultivate or capture. Farmers can increase recycling on their farms by growing more fodder, forage and cover crops and keeping animals, including fish, for manure and composting. Farmers can increase resource capacity by rehabilitating water and soil resources. These are management options farmers can pursue for more ecologically sound agriculture. Species and habitat diversification, enterprise integration and resource and habitat rehabilitation are the operational keys to ecological agriculture. In the FSA described in the next section, improvements to all these operational keys are implemented simultaneously. Simultaneous multiple intervention lies at the heart of the approach.

An Ecological Farming Systems Approach

Description

All research on farming systems involves five phases. The descriptive phase draws on the maps, transects and seasonal calendars of agroecosystems analysis (Conway, 1982; Pretty, Guijt, Thompson & Scoones. 1995). Of vital importance here is the identification of local categories of agroecological niches. Each category of niche should be defined in terms of the quality of its soil and water resources, and the species cultivated on it or supported by it. Often animals are supported by grasses cut and carried from faraway niches. This information is often presented in the form of an agroecosystems transect (Figure 17.2). Niches can also be mapped (Figure 17.4). Maps showing the relative size and location of each niche help us understand their ecological significance. Such maps can also be used to indicate access and control of natural resources. Overlaying features like access and control onto agroecosystem maps opens doors to integration of agroecosystems analysis with other analytical tools, such as those of stakeholder and gender analysis (Feldstein & Jiggins, 1994; Engel 1997; Grimble & Wellard 1996).

Diagnosis/Future Vision

The diagnostic phase usually identifies problems and elaborates a causal analysis. In this modification of FSA, we elaborate visions of future farming systems based on

Figure 17.4 Agroecosystem map from Mampong, Ghana.

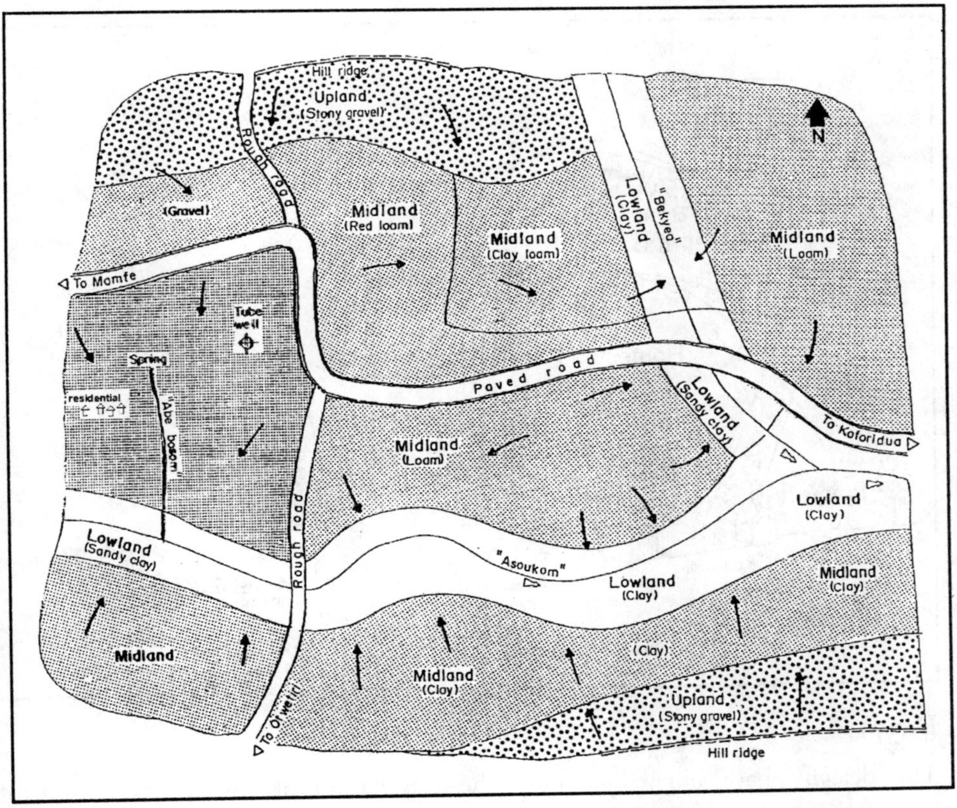

the analytical framework of ecological soundness. Farmers and researchers brainstorm ways that current farming systems can become more diversified and more integrated while degraded niches are rehabilitated. The brainstorming process is guided using bio-resource flow models (Lightfoot, Noble & Morales 1991). Bio-resource flow models (Figure 17.5) are qualitative models that show each agroecological niche, the species that are cultured or cultivated on them and the flows of bio-resources between species and niches. Farmers and researchers are challenged to envision ways in which new species can be added, degraded niches can be rehabilitated and new bio-resource flows can further integrate the farming system. The farmers each envision their own farms, so each farm is unique in its details.

Figure 17.5: Experimental bio-resource flow model from Mampong, Ghana.

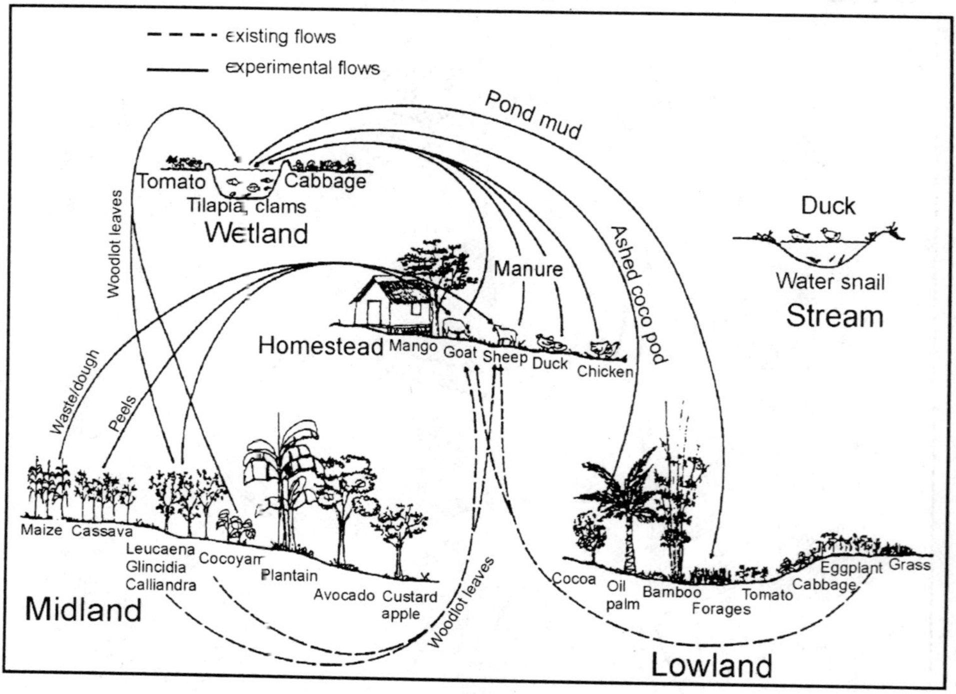

Design

The design phase normally entails the elaboration of treatments, layout of experimental plots and determination of variables to be measured. Our treatments include many enterprises. The whole farm becomes the experimental unit. Indicators of whole farm performance are evaluated. In this design phase, farmers prepare experimental bio-resource flow models as illustrated in Figure 17.5. These models superimpose the new species, new recycling flows and new resource rehabilitation techniques onto the bio-resource flow model of the existing farming system. Timeframes for these experiments vary between season and year, depending on the number of cropping seasons in the year. Every farmer has his or her own unique "experimental" model. As every farm is different, we do not have replicates that might be used in an analysis of variance. Drawing inferences from the data is best done using regression techniques and non-parametric tests. These techniques require more observations than the four replicates of an analysis of variance. Therefore, at least thirty farmers should participate. Because the experimental unit is the whole farm, there is no control plot. Comparison is made between existing farming systems and experimental farming systems. Because existing farming systems serve as the controls, special attention must be given to discounting seasonal variation. Indicators of ecological soundness consider species diversity, natural resource capacity, and bio-resource recycling flows. Economic viability is indicated by economic efficiency. These four indicators address the main questions farmers have:

- How much more profit does my farm make for every "dollar" invested?
- How much more productive is my farm?
- How much more diverse is my farm?
- How much more recycling is going on?

Testing

The testing phase sees the on-farm trials implemented. The on-farm trials proposed here belong to the farmer-managed-farmer-implemented category. Laying out plots, providing inputs and applying treatments are all relatively easy when dealing with only one intervention. Things are different when dealing with multiple simultaneous interventions (Lightfoot & Noble 1993; Lightfoot, Dalsgaard, Bimbao & Fermin, 1993). Farmers must negotiate amongst themselves and with input providers to implement their experimental farming system design. Getting all the inputs, markets, and advice they need engages many more actors than just the researcher and farmer. Extension agents are involved, as well as local NGOs and private sector representatives. Negotiation also includes the training, advice and other inputs like germplasm from researchers. Often farmers will work in groups to help each other make contour terraces, clean water courses and dig ponds. The trial must be managed by farmers themselves and they must make the decisions regarding changes to the experimental design. Some of the design elements may prove to be impractical or farmers may have been too ambitious in their plans. Learning how farmers can begin to make their farming systems more ecologically sound is vital to this research. What cannot be achieved in the first year can always be attempted in the next year and, ideally, these kinds of tests should be conducted over a number of years. Too much intervention on the part of research effectively mitigates this learning; a "hands-off" stance by researchers has the benefit of allowing many more farmers to participate in these experiments. Hence the utilization of farmer-managed, farmer-implemented trials.

Evaluation

The performance of the whole farming system is evaluated. In accordance with our analytical framework, four indicators are used:

- Economic efficiency is measured as a simple profit-cost ratio
- Resource capacity is measured as the total biomass produced from all agroecological niches.
- Species diversity is measured as the number of species cultured, captured or collected.
- Bio-resource recycling is measured as the number of flows between species and agroecological niches.

The farmers' own evaluations of the performance of their experimental farming system against their existing system are vitally important. Hence, farmers' estimates of the four indicators above are sought. At the beginning of the experiment, the bio-resource flow model of the existing farming system is examined to provide estimates of species diversity and bio-resource recycling. Farmers are asked to recall production levels from each agroecological niche and estimate farm costs and returns. The accuracy of these estimates improve as the experimental farming system becomes the existing farming system. During the period of the experiment, a season or year, farmers record on their experimental bio-resource flow models what they actually do. They record the actual number of species cultivated, the actual number of recycling flows, the actual harvests from each agroecological niche. Also recorded are the cash expenditures and sales made. At the end of the experimental round, estimates for the four indicators are calculated from the records on the experimental bio-resource flow model. The two estimates, existing and experimental, for each indicator are compared graphically in a four-way plot or kite diagram as shown in Figure 17.6. The visualization of these comparative data enables researchers and farmers to interpret changes and engage in a further round of brainstorming improvements to the farming system. These data, however, do not show how individual interventions contribute to the performance of the whole farming system. For that evaluation, farmers must return to the details recorded on their bio-resource flow models. However, the purpose of this FSA is to give farmers and researchers information on the performance of the whole farming system; information we hope will result in a farmer-led procedure to accelerate the development of ecologically sound farming systems.

Figure 17.6 Changes in farming system performance indicators from Ghana and Malawi

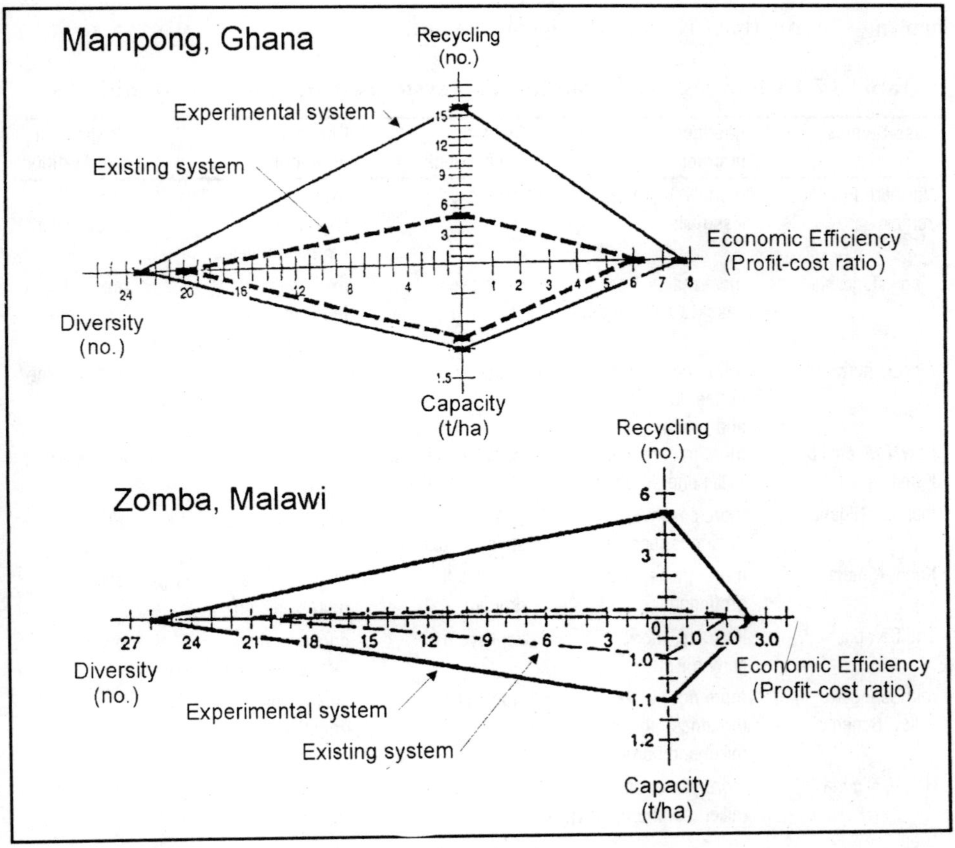

A Pattern of Achievement

In his review of fourteen case studies from Senegal, Nigeria, Gambia, Tanzania, Zambia and Kenya examining the changes in farming systems between the mid-70s and 80s, Wiggins found that farming system diversification allowed farmers to keep pace with population growth. He found, as shown in Table 17.1, that farmers have increased livestock numbers, often adding new species such as pigs, goats, donkeys and sheep. Farmers did grow more crops and have added new vegetable species, such as tomato and cabbage, to diversify their farming systems. Production methods have changed to increase yields. Farmers have adopted Green Revolution technologies such as high-yielding varieties, fertiliser and irrigation, as well as more traditional technologies like manure, inter-cropping and rotation. While these changes have kept more farmers fed and out of poverty, they have harmed soil and water resources. The evidence suggests that most African smallholders manage farming systems that are not ecologically sound.

Table 17.1 Changes in African farming systems from mid-70s to mid-80s

Case Studies	Changes in crops	Changes in Livestock	Changes in Methods	Changes in Soil Fertility
Northern Province Zambia	more HYV maize, less millet	more oxen	oxen traction, fertiliser, HYV seed, permanent fields	down, acidification
Iringa, Tanzania	more maize, less cassava and sorghum	more oxen	oxen traction, fertiliser, HYV seed, permanent fields	reduced
Mbozi, Tanzania	more maize, beans. coffee, less millet and cassava	more oxen	oxen traction, fertiliser, manure	under threat
Lower Machakos, Kenya	more maize, less millet and sorghum	large increaseoxen, sheep, goats	oxen traction, manure	fear decline
Ibarapa, Nigeria	more cassava, tomatoes, pepper	more specialization	hired tractors	no major change
Kano, Nigeria	more grains, less groundnuts,	loss due to drought,	restocking hand hoeing, more weeding	falling
The Gambia	less rice, more early millet	donkeys	donkey traction, fertiliser, manure	lower
mid Senegal valley, Senegal	more rice, tomatoes, potatoes, less millet and sorghum	stock fed	stover, pump irrigation	?
Namu, Nigeria	more yams, cassava, millet, sorghum, cowpea	pigs	added intercropping, rotation, manure and fertiliser.	?
South Nyanza, Kenya	more sugar cane			?
mid Machakos, Kenya	more maize, more cotton, fruit	pasture	development, less stock, oxen traction, manure, terracing,	not known
lower Sokoto valley, Nigeria	less rice, more sorghum, millet and cash crops	oxen	oxen traction, fertiliser, irrigation pumps	not known
Upper Embu, Kenya	more coffee, trees planted	dairy cattle	terracing, dairy cattle stall fed	not known
Upper Machakos, Kenya	more fruit, vegetables, trees planted	dairy cattle	some terracing, dairy cattle stall fed	increased

source: Wiggins, 1995

An important exception was the case of Upper Machakos in Kenya. There, Tiffen and Mortimer (1993) found that between 1930 and 1990, farmers, largely on their own, transformed their traditional agro-pastoralist systems to intensive mixed farming without degrading the natural resource base. Farmers achieved this by diversifying into cash cropping and horticulture, and integrating trees and stall-fed dairy cattle along with land terracing into their farming system. Under these integrated mixed farming systems, soil fertility demonstrably increased and agricultural output per head rose from 0.3 to 1.1 tons/ha. Unhappily, this remarkable success story rarely finds parallels elsewhere. Indeed, it was the one exception in Wiggins' fourteen cases.

Two cases, one from Ghana and one from Malawi, suggest that if more researchers and farmers pursued an ecological FSA, things might be different. In both countries, the participating farmers were smallholders initially running a typical mixed farm with livestock, trees, vegetables, cereals and root crops for home consumption and sale. A few goats, sheep and chickens provided meat. Recycling was limited to collecting various forages, crop residues and leaves for the livestock. Manure was hardly used at all. There was, of course, variation between the farms in both countries, but the pattern was common. Using bio-resource flow models, farmers and researchers brainstormed more ecologically sound experimental farming systems. On-farm trials were conducted with farmer groups, following the methods for testing simultaneous multiple interventions described here. Over a period of 18 months, the farms changed dramatically. While each farm was different, all variously rehabilitated water resources, established vegetable plots, integrated fish (and, in few cases, ducks) into their farming systems. Much more recycling occurred. Previously neglected wetland areas – dambos, in the case of Malawi – were rehabilitated by improving water flow and impounding water in small ponds. These ponds provided sufficient water to stock fish and irrigate newly established vegetable plots. Manure was recycled to the pond and vegetable plot, as were cocoyam and leucaena leaves. In Malawi, pond mud was recycled to the vegetable plots to substitute for chemical fertilizer. With savings on external inputs, more products going to market, and more internal recycling of nutrients, overall farming system performance changed dramatically.

Changes in farming system performance were evaluated. In just 18 months, all indicators improved, as Figure 17.6 shows. In Ghana, resource rehabilitation and increases in recycling from 5 to 16 flows and species diversity from 20 to 23 brought modest increases in biomass production and more impressive increases in economic efficiency (Prein, Ofori & Lightfoot, 1995). Similar results were reported in Malawi (Lightfoot & Pullin, 1995; Brummett & Noble, 1995). Again, in a little over 18 months, improvements had been made in all indicators. Recycling flows increased from zero to 5, and species diversity from 21 to 26. The rehabilitation of water resources for culturing fish and growing vegetables in dambo lands resulted in modest rises in economic efficiency and biomass production. As there was a drought in Malawi over this period, it is surprising to see any increases at all.

The integrated mixed farming systems that emerged in Malawi and Ghana were not only more ecologically sound, they were also more profitable. As indicated in Table 17.2, income increases from rehabilitating water resources in lowlands and wetlands varied from as little as US$32 in Malawi to as much as US$202 in Ghana. Rehabilitation costs, which comprised mainly labour for clearing or diverting water courses, impounding water or digging ponds, were low at more or less US$37. This, however, did mean that farmers in Malawi showed a loss in the first year. The balance of all the interventions made in the Malawi case provided a modest gain of 4% on gross income and 15% on net income. This was because pond mud and compost substituted for chemical fertilizer in the wetland dimba vegetable gardens, allowing a considerable savings of US$80. Much more dramatic increases in gross and net incomes were realized from all the changes made in the Ghana case, where gross income rose from US$1,686 to US$2,524. (Lightfoot & Pullin, 1995).

Table 17.2 Changes in Economic Performance on Case Study Farms in Ghana and Malawi

GHANA	Existing Farming System				Experimental Farming System			
	Gross Income	Total Cost	Net Income	Net Cash Income	Gross Income	Total Cost	Net Income	Net Cash Income
Lowland	921	27	894	600	1481	63	1418	889
Upland	566	200	366	430	819	135	684	638
Homestead	167	12	155	132	22	47	-25	9
Fishpond	0	0	0	0	202	36	166	192
River	32	0	32	0	0	0	0	0
Farm	**1686**	**239**	**1447**	**1135**	**2524**	**281**	**2243**	**1728**
% Change					50%	18%	55%	52%

MALAWI	Existing Farming System				Experimental Farming System			
	Gross Income	Total Cost	Net Income	Net Cash Income	Gross Income	Total Cost	Net Income	Net Cash Income
Homestead	142	14	128	78	108	18	90	20
Dimba	271	106	165	112	296	26	270	173
Munda	205	96	109	-15	205	97	108	-34
Dambo	0	0	0	0 3	2	38	-6	23
Farm	**618**	**216**	**402**	**175**	**641**	**179**	**462**	**182**
% Change					4%	-17%	15%	4%

While Malawi's data suggests poor performance, the farmers did not see it that way. They felt that making any money at all in a drought year was a considerable achievement. Moreover, the farmers were impressed with other benefits, as indicated in Table 17.3. In their perception, household nutrition was improved through more regular rice and fish consumption. They saved money by using composts and pond mud instead of chemical fertilisers to maintain the soil fertility of their vegetable plots. Perhaps most critical of all was extending water supplies for domestic use, livestock watering and vegetable irrigation. During the 1991/92 drought, these households managed to feed themselves without losing their savings, a considerable advantage for resource-poor farmers (Lightfoot & Noble, 1993)

Table 17.3 Malawi farmer perception of the benefits of integrated farming systems

Existing Farming System (1988/89)	Experimental Farming System (1990/91)
Marginal wetland unutilized	Marginal land brought into productive use
No integration between resource systems or farm enterprises	Ponds serve as a focal point for direct or indirect links between resource systems
Crop residues not recycled	Crop residues such as maize bran and green leaf waste used as pond input
Water shortage for vegetable garden in late dry season	Ponds placed adjacent to garden provide water for irrigating vegetables Average dimba (vegetable garden) earnings increased from US$82 to US$112/year
Households reliant on uncertain water supplies	Households use pond as water catchment for domestic use and watering livestock
Reliant on fertilizer for vegetable garden or overutilize exhausted soils	Use ponds as processing unit for converting low quality crop waste into fertile mud for transfer to garden; reduced fertilizer use
Buy fish; rarely eat fish	Ready supply of fish for household consumption; rarely buy fish. Average fish harvest values: US$22 (1 harvest), US$45 (2 harvests)
Marginal wetland does not provide food and income for household	Conversion to ponds provides food and income (Average income from integrated rice-fish = US$76/year)
Rice either not grown or only one crop per year	Rice-fish ponds provide two crops of rice per year; rice grown for first time
Average annual income before integration (1990) = US$155	Average annual income after integration (1992) = US$235. A year that saw a nationwide drought and a currency devaluation of 20%

source: Lightfoot and Noble, 1993

These results should not be seen as a blanket validation of the ecological FSA. The on-farm trials in Malawi and Ghana engaged only some thirty farmers for a few years. The farms were located in fairly favorable agroecological conditions of transition zones between savanna grasslands and forests. Fairly heavy external assistance from research and extension was provided. These results do, however, add to the growing body of evidence that increasing species diversity, recycling nutrients, and rehabilitating soil and water resources can secure marketable surpluses at economically attractive rates and in an ecologically sound manner for smallholders. Similar results have been found in Asia. In their ecological analysis of four smallholder farms ranging from monocrop rice farms to highly integrated and diversified rice farms, Dalsgaard and Oficial (1996) found that the integrated farm was more ecologically sound than the monocropping farm. Their findings support the notion that a farm with higher scores in the attributes discussed is indeed more ecologically sound. The FAO has reported that these types of farming systems are also economically viable and concluded that net incomes rose as the farming system became more diverse and integrated (FAO/RAP, 1996). We do not know whether the ecological FSA approach described here can engage researchers and farmers in a much more rapid spread of ecologically sound farming systems. A much wider and more rigorous use of this farming systems approach is needed to confirm its utility.

The Challenge for Researchers

The application of the ecological FSA outlined here has some serious implications for national agricultural research organizations. It is almost a pre-condition that they have a mandate for farming system research and policies embracing participatory research. National research organizations with farming system research teams based at research stations or local offices are perhaps in the best position to move quickly. Still, in most cases, they will have to wrestle with the switch from diagnosis to vision, from single intervention trials to multiple simultaneous intervention trials, and from research to multiple service provision.

From diagnosis to future vision

Diagnostic procedures and thinking have become deeply embedded in research organizations. They are easy to get from farmers and fit nicely with research disciplines and commodity mandates. Only a few researchers have noticed that the endless round of diagnostic surveys throwing up the same set of apparently intractable problems leaves farmers depressed and drained of energy. Fewer still have experienced the enthusiastic and energizing reaction farmers give to the subject of future visions. The bio-resource flow models visualized by farmers communicate well among farmers and with other stakeholders. Local people really get excited by them. Getting researchers to change from the known comforts of diagnosis to the unknown of future visions will not be easy. Nevertheless, it is not the solving of present-day problems but the striving towards future visions of ecologically sound agriculture that drives on-farm experimentation in this approach. Farmers need outsiders to enhance visioning exercises with new knowledge. Multiple sources of information and innovative ideas are needed. Often the researchers can connect farmers to new sources of information. In turn, researchers also benefit from these exercises. What farmers cannot realize and what is not known provide insights into research agenda development. Over time, these exercises can lay a foundation for effective, bottom-up demand on research and development services – something many national research organizations desire.

From single intervention to simultaneous multiple interventions

Moving from testing single interventions to many interventions all at the same time requires big compromises in experimental control, precision and accuracy. The change in analysis unit, from the plot to the whole farm, effectively removes the possibility of having a control plot. Comparisons with neighbouring farmers or with past performance can never be as precise as control plots. By taking the farm as the unit of analysis, contributions to performance by particular commodities can often not be determined. Worse still, in the eyes of some, are the compromises in yield that must be made for the system to function in ecological balance. One extreme instance of such compromise occurred in the Malawi case described here. At the height of the drought, the water in the fishponds dried up. Farmers harvested their fish early at extremely low yields and planted vegetables into the drying pond. The vegetables did very well. Compromising on fish yield but producing vegetables improved overall farm performance, but the aquaculture bosses were not impressed. A further compromise for some will be the necessary relinquishing of decision-making power over experimental design. Farmer-managed, farmer-implemented trials mean farmers decide on management. Given all these compromises, many researchers will not want to undertake such on-farm trials, despite the far greater impact of multiple simultaneous intervention on the performance of a whole farming system. Even doubling the yield of one subsistence crop on a smallholder farm, where it is planted to a small area, has small impact on household income or farming system output. The really large changes needed to take farmers out of poverty and rehabilitate the many degraded natural resources can only be effected by multiple simultaneous interventions.

From research team to wide array of actors in experiments

An ecological FSA requires that researchers work with many more people than just the farmers. The cosy mode of a research team working with a handful of "cooperating" farmers will be hard to break. The reality of spreading limited resources more thinly is often enough for farmers to start worrying. But for reasons of experimental rigour, greater numbers of farmers must be reached by the same research team. Covering more farmers would undoubtedly be easier if farmers were better organized. It helps to work in collaboration with local NGOs mandated to organize farmers. But it is in the experimental design and testing phase that farmers need many new partners. In design, farmers need multiple sources of information from outside their own pool of knowledge. New outside information on restoration technologies, biodiversity for money-making options, as well as input substitution and bioresource recycling greatly enhance experimental designs. In testing, farmers must negotiate for the necessary inputs, advice, training, credit and markets to realize their future vision. Providing opportunities for the relevant actors to come together will present researchers with the challenge of learning how to work together and sort out a complementary set of roles. If this is done well, then ending the research component should not slow down the pace of change on the farm.

National agricultural research organizations taking up such an ecological farming systems approach will not find themselves alone. There is a large and growing literature on systems participatory approaches to ecosystems management (Bawden, 1991; Finger & Verlaan, 1995; Pretty, 1995; Daniels & Walker, 1996; Ison, Maiteny & Carr 1997). In their recent edited volume, Roling and Wagemakers bring together experiences from the Netherlands, Germany, Indonesia, Australia, and the USA. All of these cases stress the importance of farmer participation, multiple stakeholder negotiation for concerted action, multiple sources of information and continuous learning processes. To facilitate these participatory learning processes, researchers need to form coalitions and alliances or platforms with extension, NGOs and private sector providers of agricultural services (Roling, 1994; Biggs, 1997). Such alliances will begin to provide farmers with a wider range of information sources. The need for a special knowledge network for ecological agriculture has been identified for some time now (Roling & Jiggins 1996). Indeed, participatory methods to promote the development and understanding of such farmer communication networks have been developed (FAO, 1995; Ramirez 1997). Much of the methodological development to support the ecological FSA proposed here has been in the realm of participatory research. In the end, as Chambers says, "Whose reality counts?" (Chambers 1997). Getting national research organizations to recognize this and engage in a learning process with farmers for the development of ecologically sound agriculture is the challenge.

CONCLUSION

Farming systems "my way" has been an enduring feature of the farming systems approach to agricultural research. The ability to be "reworked" to suit many different circumstances is both a weakness and a strength. The multiplicity of interpretations and applications presents a confusing picture to just about everyone. This has not helped the promotion or the sustained financial support to farming systems research. On the other hand, there are strengths in such versatility. It has given innovators room to fashion their version of the FSA towards their own particular ends. This paper has presented an ecological farming systems approach drawing from the methods of agroecosystems analysis and farmer participatory research. The approach stresses the involvement of farmers in visioning the ecological changes to be introduced, the study of whole farms as the unit of analysis and the introduction of multiple simultaneous changes. This approach presents challenges to researchers, but evidence of its success in Africa and elsewhere continues to accumulate.

REFERENCES

Altieri. M.A. 1989: *Agro-ecology: The Science of Sustainable Agriculture.* Westview Press, Boulder. Colorado, USA 433p.

Bawden, R. 1991: Systems Thinking and Practice in Agriculture. *Journal of Dairy Science* (74):2362- 2373.

Biggs, S 1997: Livelihood, Coping and Influencing Strategies of Rural Development Personnel. *Project Appraisal* 12 (2):101-106.

Brummett, R.E. and Noble, R. 1995: *Aquaculture for African Small-holders.* ICLARM Technical Report 46. 69p.

Byerlee. D., Triomphe, B. and Sebillotte, M. 1991: Integrating Agronomic and Economic Perspectives in the Diagnostic Stage of On-Farm Research. *Experimental Agriculture* 27:95-114.

Chambers, R. 1997: *Whose Reality Counts? Putting the First Last* London: IT Publications.

Collinson. M.P. 1981: A Low Cost Approach to Understanding Small Farmers. *Agricultural Administration* 8:433-450.

Conway G. R. 1982: A Guide to Agroecosystem Analysis. In *Tambon and Village Agricultural Systems in Northeast Thailand.* KKU Ford Cropping Systems Project, Khon Kaen University, Thailand.

Conway G. R. 1985: Agroecosystem Analysis. *Agricultural Administration* (20): 31-55.

Dalsgaard, J.P.T., Lightfoot, C. and Christensen, V. 1995: Towards quantification of ecological sustainability in farming systems analysis. *Ecological Engineering* (4):181-189.

Dalsgaard J.P.T. and Oficial, R.T. 1996: Agroecological systems analysis of four smallholder rice farm scenarios: modeling, quantifying and comparing ecological attributes. In *An ecological modeling approach towards the determination of sustainability in farming systems*. PhD Thesis, J.P.T. Dalsgaard, Royal Agricultural and Veterinary University, Copenhagen, Denmark.

Daniels, S.E. and Walker, G.B. 1996: Collaborative Learning: Improving Public Deliberation in Ecosystem-Based Management *Environmental Impact Assessment Review* (16): 71-102.

Edwards, P., Pullin, R.S.V., Gartner, J.A. 1988: Research and education for the development of integrated crop-livestock-fish farming systems in the tropics. ICLARM *Stud. Rev.* (16): Manila, Philippines. 53p.

Ellis. F. 1998: Livelihood Diversification and Sustainable Rural Livelihoods. pp 53-66. In D. Carney (ed.) Sustainable Rural Livelihoods. London, UK: DFID. 209p.

Engel, P. 1997: The Social Organization of Innovation. A Focus on Stakeholder Interaction. In: Engel, P. and Salomon, M. *Facilitating Innovations for Development*. Amsterdam, Netherlands: KIT, STOAS and CTA, 239p.

FAO. 1995: *Understanding Farmers Communication Networks: An Experience in the Philippines.* Communications for Development Case Study 14. FAO, Rome, Italy. 74p.

FAO/RAP. 1996: *Low External Input Sustainable Agriculture in Selected Countries of Asia.* (Regional Workshop Report with Case Studies. RAP publication 1996/19. FAO/RAP. Thailand.)

Feldstein H.S. and Jiggins, J. 1994: *Tools for the Field: Methodologies Handbook for Gender Analysis in Agriculture.* West Hartford, Connecticut: Kumarian Press. 270p.

Finger M. and Verlaan, P. 1995: Learning Our Way Out: A Conceptual Framework for Social-Environmental Learning. *World Development*. 23 (3) pp. 503-513.

Grimble, R. and Wellard, K. 1996: Stakeholder Methodologies in Natural Resource Management: a Review of Principles, Contexts, Experiences and Opportunities. *Agricultural Systems.* Vol 65, no 2. pp 173-193.

Hildebrand. P.E. 1986: (ed.) *Perspectives on Farming Systems Research and Extension.* Boulder, Colorado: Lynne Rienner Publishers.167p.

Ison, R.L., Maiteny, P.T. and Carr, S. 1997: Systems Methodologies for Sustainable Natural Resources Research and Development. *Agricultural Systems*. Vol 55. No.2. pp:257-272.

Jorgensen S.E. 1992: *Integration of Ecosystem Theories: A pattern*. Dordrecht, Netherlands: Kluwer Academic Publishers.

Kenmore, P. E. 1991: *Indonesia's Integrated Pest Management: A Model for Asian Rice Farmers' Environmental Conservation, Production, and Profits*. Manila, Philippines: FAO. 56p.

Lightfoot, C. and Noble, R. 1993: A Participatory Experiment in Sustainable Agriculture. *Journal for Farming Systems Research and Extension*. 4(1): 11-34.

Lightfoot, C. and Pullin, R.S.V. 1995: An integrated resources management approach to the development of integrated aquaculture farming systems, p145-168. In J.J. Symoens and J.C.Micha (eds.) *The management of integrated freshwater agro-piscicultural ecosystems in tropical areas.* Brussels: CTA, Royal Academy of Overseas Sciences, 587p.

Lightfoot, C., Dalsgaard, J.P., Bimbao, M., and Fermin, F. 1993: Farmer Participatory Procedures for Managing and Monitoring Sustainable Farming Systems. *Journal of the Asian Farming Systems Association* 2 (1): 67-87.

Lightfoot, C. 1990: Integrated Route to Sustainable Farming. *International Agricultural Development.* 10(2): 9-11.

Lightfoot, C., Noble, R and Morales R. 1991: *Training Resource Book on a Participatory Method for Modelling* Bioresource Flows. ICLARM Educ. Ser. 14. Manila: ICLARM,. 30p.

Mollison, B. 1993: *Introduction to Permaculture*. Tyalgum. Australia: Tagari Publications. 198p.

Norman, D.W. 1978: Farming systems research to improve the livelihood of small farmers. *American Journal of Agricultural Economics* 60:813-818.

Norman, D.W. and Lightfoot, C. 1992: A Formative Evaluation of the FAO/SIDA Farming Systems Program. Rome, Italy: FAO.

Odum E.P. 1969: The strategy of ecosystem development. *Science* 104: 262-270.

Prein, M., Ofori, J.K. and Lightfoot, C. 1995: *Research for the Future Development of Aquaculture in Ghana.* ICLARM Conf. Proc. 42. Manila, Philippines.

Prein, M., Lightfoot, C. and Ofori, J.K. 1996: Partizipative Methoden in der Einfuhrung von Semi-Intensiven Aquakultursystemen in die Traditionelle Ressourcenberwirtschaftung. In *Integierte Aquakulture.* H. Franzen and H. Rosenthal (eds.) ATSAF, Bonn, Germany.

Pretty, J., Guijt, I., Thompson, J. and Scoones, I. 1995: *Participatory Learning and Action: A Trainer's Guide.* IIED Participatory Methodology Series. London: IIED. 267p.

Pretty J.N. 1995: *Regenerating Agriculture: Policies and Practices for Sustainability and Self Reliance.* London: Earthscan.

Raintree, J.B. 1984: Designing agroforestry systems for rural development: ICRAF'S D & D approach. International Council for Research in Agroforestry.(ICRAF) Nairobi, Kenya. 19p.

Rambo, T.A. 1984: Human Ecology Research by Social Scientists on Tropical Agroecosystems. pp: 39-60 In T.A. Rambo and P.E. Sajise (eds.) *An Introduction to Human Ecology Research on Agricultural Systems in Southeast Asia*. Los Banos, Philippines: Univ. of the Philippines.

Ramirez, R. 1997: *Understanding Farmers' Communication Networks: Combining PRA with Agricultural Knowledge Systems Analysis*. IIED Gatekeeper Series No.66. IIED. London, UK.

Roling. N. 1994: Platforms for decision making about ecosystems. In Fresco L, Stroosnijder L, Bouma J, van Keulen H. (eds.) *The Future of the Land: Mobilizing and Integrating Knowledge for Land Use Options.* New York, USA: John Wiley.

Roling N. and Jiggins, J. 1996: The Ecological Knowledge System. In *Proceedings 2nd European Symposium on Rural and Farming Systems Research*. Granada, Spain.

Roling, N. and Wagemakers, J. 1998: *Facilitating Sustainable Agriculture* Cambridge, U.K.: Cambridge University Press.

Sajise, P.E. 1986: Agroecosystem Analysis: The SUAN Approach. *Proceedings 3rd SUAN-EAPI Regional Symposium on Agroecosystem Research*. Chaing Mai University, Thailand October 1996. East West Environment & Policy Institute, Hawaii, USA.

Tiffen M., and Mortimer, M. 1992: Environment, Population Growth and Production Intensity: A Case of Machakos, Kenya. *Development Policy Review* (10) 359-387.

Wiggins, S. 1995: Change in African Farming Systems between the mid 1970's and the mid 1980's. *Journal of International Development*: 7(6):807-848.

CHAPTER 18

THE ROLE OF A FISH POND IN INTEGRATED RESOURCE MANAGEMENT

by Fredson J.K Chikafumbwa, Randall F. Brummett and Reg P. Noble

THE INTEGRATED RESOURCE MANAGEMENT APPROACH

Farmers adopt most production practices as a means of providing food security and income for their families. A major aim of development research, whether in aquaculture, forestry, livestock or crops, should be to create sustainable improvements in the food security of resource-poor farmers. Lightfoot and Noble (1993) comment that most development research is commodity oriented and does not address the needs of farmers managing their land for maximum food security rather than income.

More than 80% of Malawi's farmers are resource-poor smallholders. Fish ponds linked with other farming activities, such as rice or vegetable production, can generate significant improvements for the farming household and its environment. This suggests that researchers should take a broader look at the role fish ponds could play on small farms. The Integrated Resource Management (IRM) approach to developing integrated aquaculture seeks to transform existing farming systems and water resources into an integrated aquaculture farming system.

In Malawi, aquaculture has typically been treated as a stand-alone enterprise in the farming system. For example, many aquaculture projects have presented farmers with technological packages designed to function only as an independent commercial operation (GOPA 1987). In contrast, the IRM approach seeks to integrate aquaculture into farming systems, using a farmer participatory research protocol, guided by ecological and economic modelling and supported by socioeconomic studies. This approach is designed to ensure that the systems are economically and socially sustainable for resource-poor smallholders. IRM seeks to rehabilitate water resources for plants, livestock and fish, as well as increase incomes and food security. It also builds knowledge and skills for sustainable use of resources (Table 18.1).

Table 18.1 Comparison of conventional and IRM approach to integrated aquaculture research and development

	CONVENTIONAL APPROACH	IRM APPROACH
1.	High external input	Low external input
2.	Fish production objective	Sustainable resource management objective
3.	Pond as unit of analysis	Whole farm as unit of analysis
4.	Commodity perspective	System perspective
5.	Researcher designed and evaluated	Farmer-designed and evaluated
6.	Improving yields of fish farmers	Improving whole farm efficiency

Source: Lightfoot and Pullin 1994

Steps In The IRM Farmer Participatory Process

The IRM approach involves farmers in the rehabilitation of their water resources and the integration of aquaculture into on-going farming activities. The process involves four main steps:

- **Identification of the natural resource types in the area of concern – watershed, catchment, village or farm:** Local indigenous knowledge is tapped by asking farmers to describe water, land and agricultural resources available in the area. Farmers are then engaged in a discussion of how these resources interact, either positively or negatively. Topological transects and careful categorization of natural resource types provides a framework for comparing systems across regions (Conway, McCracken & Pretty, 1987; Lightfoot, Axinn, Botrall & Conway, 1989; Khan, Lai, Quadir & Iqbal Ali, 1990).
- **Modelling and designing experimental bioresource flows for an integrated aquaculture farming system:** These models bring farmers and scientists together to develop ways to rehabilitate water resources and integrate aquaculture and other enterprises into on-going farming systems. Interested farmers are encouraged to make drawings of their own farms. The models provide the basis for farmers and scientists to discuss how technologies for water resource rehabilitation and aquaculture integration can fit into their systems. Through these discussions, four outputs emerge:
  - ideas for immediate adoption
  - ideas for on-farm experiments
  - ideas for on-station research
  - feedback for policy makers.

- **Monitoring by farmers:** Farmers themselves monitor the bioresource flows, the natural resource rehabilitation and the enterprise additions that are introduced. The data collected is qualitative and researchers may want to supplement these data with quantitative data of their own. The process of matching farmer, qualitative valuation and scientist's quantitative valuation, provides useful insights into improving the management of natural resources.
- **Evaluation of economic and sustainability performance:** Evaluating the performance of an integrated aquaculture system extends to the examination of sustainability indicators for the whole farm system. Four-way plots, or kite diagrams, of species diversity, bioresource recycling, productive capacity of natural resource base and economic efficiency are created. Species diversity simply counts the number of species used. The number of bioresource flows provides the value for resource recycling. The biomass production coming from natural resource types is used to estimate their capacity to produce. Economic efficiency is estimated using profit-cost ratios. Evaluation combines data from farmer profiles, transects and the monitoring diagram.

Fish Ponds in Integrated Resource Management

Using open days at the National Aquaculture Center, Domasi and Zomba farmers were exposed to various crop-fish integrations. Between 1990 and 1993, farmers from Zomba, Mulanje and Mangochi were invited to different open days.

In December 1990, 17 fish farmers from Zomba, where rice is widely grown, were shown rice and fish grown together during the dry months of the year. These farmers were excited – this was the first time they had seen rice and fish managed together, although most of them had fishponds adjacent to rice paddies. They asked a lot of questions about rice-fish integration. They also criticized the fish refuge position in the rice field in relation to the slope of the field, and they gave advice on the proper setting. To capture these suggestions, the farmers led a workshop discussing the demonstration and its relevance to their environment. They were encouraged to draw what they suggested could be the possible arrangements. The result was a series of different rice-fish system designs, drawn depending on what farmers thought was relevant to their farming environment. This suggested that they understood the technology presented to them, even though this was their first experience of the integrated rice-fish system. It also demonstrated that the farmers were capable of contributing effective ideas in an area where they had no experience.

The resulting interest in integration was intense and many farmers altered their farming environment too quickly for the researchers to capture the process. However, five farmers who were relatively new to aquaculture systems in the Zomba district were asked to make drawings of their farms before and after incorporation of fish ponds (Figures 18.1 and 18.2)

Figure 18.1 Resource flows prior to introduction of fish ponds

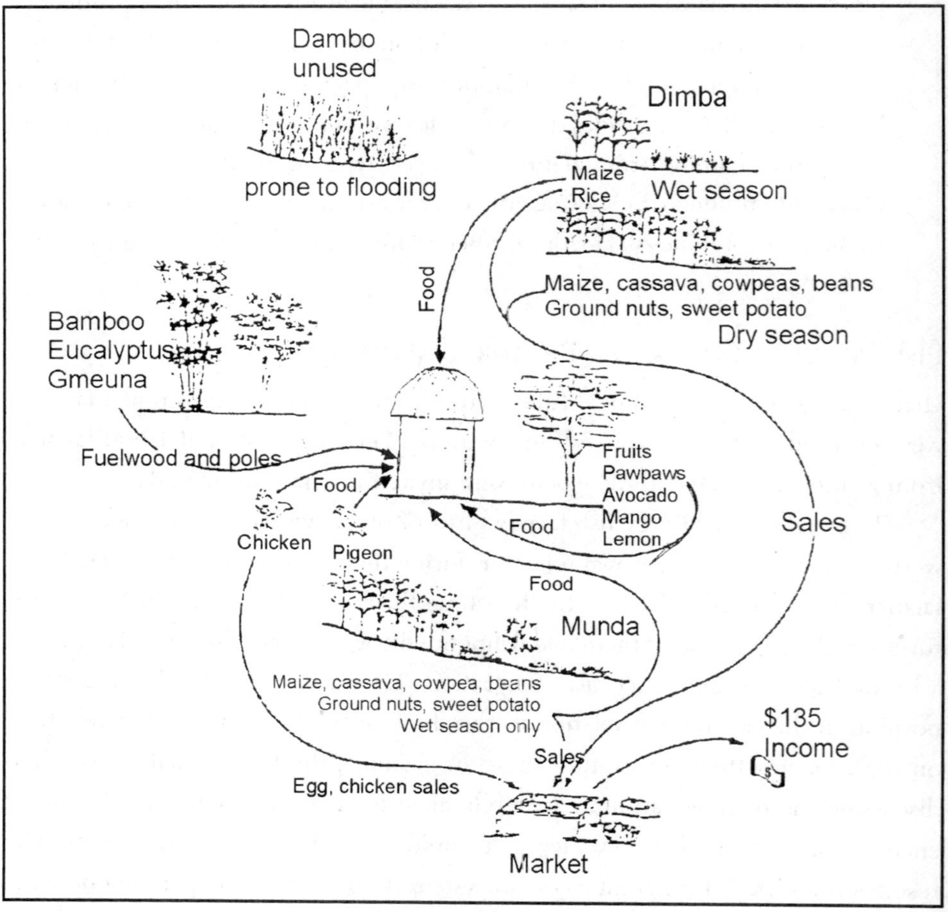

Figure 18.2 Resource flows after incorporation of fish ponds

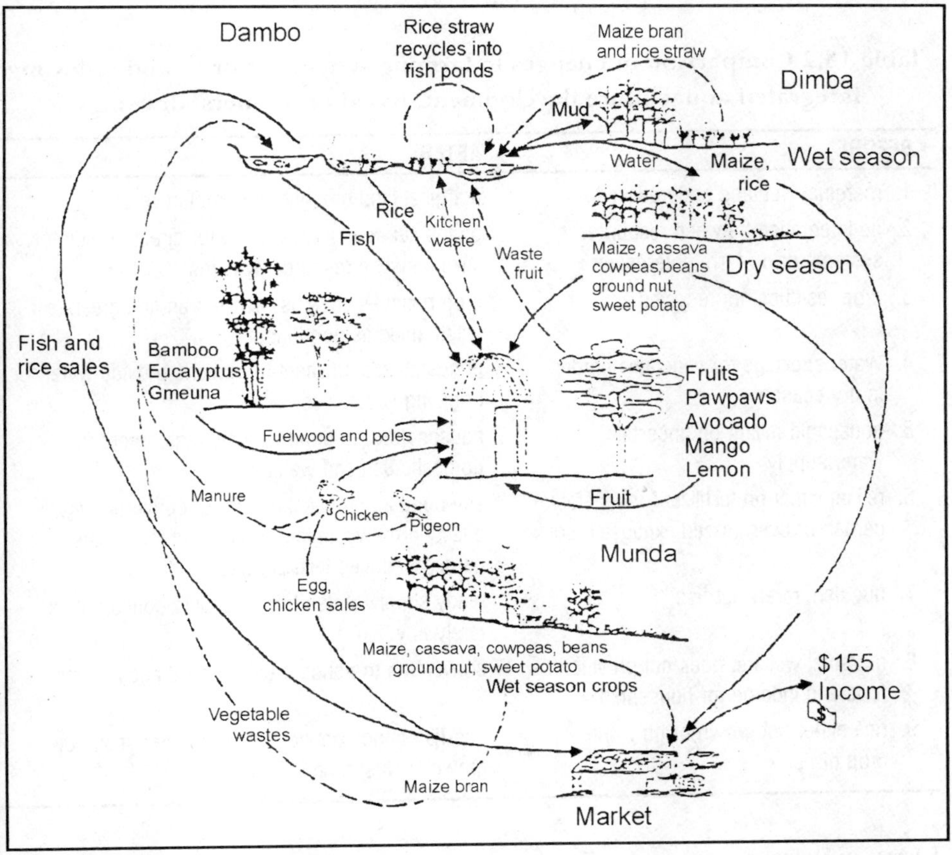

The incorporation of fish ponds brought significant alterations to the management of their farms. Increased linkages were created between their bioresource elements, with rice-fish ponds acting as the focus of new resource management strategies.

Table 18.2 summarizes the information obtained from the farm drawings. In every case, the presence of the ponds created linkages through recycling of farm residues. Farmers also commented that having ponds reduced their need to buy fish and the latter were often used for reciprocal exchange of goods and services provided by neighbours. Marginal land was brought into production for rice or vegetables and for fish. Household income was increased directly through the sale of fish and crops from the revitalized dambo resource system. These patterns were common to four of the farms sampled in Zomba West and each case involved bioresource linkages to adjacent dimba vegetable gardens with flows operating in both directions between ponds and dimba systems.

Table 18.2 Comparison of changes in farming systems prior to and following integrated aquaculture development, based on farmers' drawings.

BEFORE	AFTER
1. marginal wet land underutilised	marginal land brought into productive use
2. no integration between resource systems or on-farm enterprises	ponds serve as a focal point for direct or indirect links between resource systems
3. crop residues not recycled	crop residues such as maize, bran and green leaf waste used as pond input
4. Water shortage for vegetable garden in dry season	ponds placed adjacent to gardens provide water for irrigating vegetables
5. household reliant on uncertain water supply	household uses pond as water catchment for domestic use and watering
6. reliant either on fertilizer for vegetable garden or over utilized exhausted soils	use pond as processing unit for converting low quality crop waste into fertile mud for transfer to garden; reduced fertilizer use
7. buy fish; rarely eat fish	ready supply of fish for household consumption; rarely buy fish
8. marginal wetland does not provide food and income for household	conversion to ponds provides food and income
9. rice either not grown or only one crop of rice	rice-fish ponds provide two crops per year, rice grown for first time

Farmer Participation in Technology Development and Transfer: Rice-Fish Example

The greatest potential for fish culture appears to lie with farmers who are already engaged in the production of aquaculture-compatible crops (e.g., rice) and livestock (Brooks & Maluwa, 1993, Chikafumbwa 1994). The groups from Zomba, Mulanje and Mangochi reacted quite differently to their exposure to the integrated rice-fish farming system. Within one year, 11 of the Zomba farmers who attended the open day had started rice-fish integration. More remarkably, within two years, at least 40 other farmers were practising rice-fish integration – farmers who had not attended an open day, had never been to the National Aquaculture Center and were not contacted by the extension worker. The designs varied among the farmers. Rice-fish integration was gradually spreading in Zomba through farmer-to-farmer dissemination.

In contrast, fish farmers from Mulanje and Mangochi, where rice was seldom grown, were not as excited as their counterparts from Zomba. Rice-fish integration looked new to them and they were unable to pinpoint the weaknesses in the set-up. The questions they asked were more concerned with learning about rice-fish integration than with analysing the demonstration system. By 1993, only three of the ten farmers from Mulanje and three of the eight farmers from Mangochi (Namwera) had started rice-fish integration. No farmers who had not attended the open day started rice-fish integration.

Conclusions: Farmers Response to New Technology

The adoption of rice-fish integration indicated two different responses to the technology. In Zomba, farmers who were growing both rice and fish on adjacent farms were very excited. They understood the integration and visualized how it could fit into their farming systems without any major changes or external inputs. They were able to extend the technology to other farmers because their ideas were respected, which gave them confidence in teaching the technology. Neighbouring farmers were satisfied with the technology. The introduction of the technology provided room for flexibility, which allowed for different integrated rice-fish designs. For farmers in Mulanje and Mangochi, however, the technology required changes in their existing farming systems. Both rice and fish had to be introduced into most households. This led to a low adoption rate of the new concept. It was difficult for the farmers who attended the open days to explain the technology properly to other farmers.

The response of the farmers to rice-fish integration suggests that researchers need to understand the existing resources and farming systems of an area to which the new technologies will be introduced. Farmers have expert knowledge of their local resources and they have the skills to manage a range of innovations which can be integrated into their farming systems. Farmers participating in research design and implementation understand the management practices and results of the technology better than those who don't. Successful results will make farmers implement the technology and help them spread the ideas to other farmers with confidence. Researchers and extension workers need to realize that farmers will easily adopt technology when they know it will work, have perceived the benefits and understand the concepts.

REFERENCES

Brooks, C. and Maluwa, A. O. 1993: *Central and Northern Regions Fish Farming Project.* Research findings 1990-1993. Mzuzu, Malawi: Malawi Fisheries Department.

Chikafumbwa, F. J. 1994: Farmer Participation in Technology Development and Transfer in Malawi: a Rice-Fish Example. In R.E. Brummett (ed). *Aquaculture Policy Options for Integrated Resource Management in SubSaharan Africa.* ICLARM Conf. Proc. 46. Manila, Philippines: International Center for Living Aquatic Resources Management.

Conway, G. R., McCracken, J. A. and Pretty, J. N. 1987: *Training Notes for Agroecosystem Analysis and Rapid Rural Appraisal.* International Institute for Environmental and Development. 2nd edition, 60 pp.

GOPA, 1987: *Malawi Fisheries Development Strategy Study.* 41 pp. GOPA Consultants, 6380 Bad Homberg, Hindenbugring 18, Federal Republic of Germany.

Khan, A. M., Lai, C. K., Quadir, R. and Iqbal Ali, S. 1990: *Handbook on Land Use Mapping Resources and Sketch Mapping Techniques for Bangladesh.* 46 pp. Dhaka, Bangladesh: Ford Foundation.

Lightfoot, C. and Pullin, R.S.V. 1994: Why an Integrated Resource Management Approach? In R.E. Brummett (ed.), *Aquaculture Policy Options for Intergrated Resource Management in SubSaharan Africa.* ICLARM Conf. Proc.

Lightfoot, C., Axinn, N., Sigh V. P., Bottrall, A. and Conway, G. 1989: *Training Resource Book for Agro-Ecosystem Mapping.* 40 pp. Los Banos, Philippine: IRRI.

Lightfoot, C. and Noble, R. P. 1993: A Participatory Experiment in Sustainable Agriculture. *Journal for Farming Systems Research-Extension* 4 (I): 11-34.

Lightfoot, C. and Pullin, R. S. V. 1995: An Integrated Resource Management Approach to Development of Integrated Aquaculture Farming Systems, pp. 145-167. In J.J. Symoens and J.C. Micha (eds.) *The Management of Integrated Freshwater Agro-Piscicultural Ecosystems in Tropical Areas.* Brussels, Belgium: Technical Centre for Agricultural and Rural Co-operation (CTA) and Royal Academy of Overseas Science.

CHAPTER 19

FARMER-SCIENTIST RESEARCH PARTNERSHIPS FOR AQUACULTURE TECHNOLOGY DEVELOPMENT IN MALAWI[1]

by Randall E. Brummett and Fredson J.K. Chikafumbwa

INTRODUCTION

Increasing population pressure in sub-Saharan Africa has led to over-utilization of land and a subsequent decline in actual and potential agricultural productivity. Putting more land under cultivation will only exacerbate the decline of environmental quality (Brummett, 1995). New methods for increasing the efficiency, productivity and attainability of small-scale farming systems must be found. A logical first step is the development of improved methods for problem analysis and technology development aimed specifically at the smallholder. The International Centre for Living Aquatic Resources Management (ICLARM) has used a realistic alternative to traditional technology development and transfer to integrate pond fish culture into Malawi's low-input farming systems.

More than 80% of the people of Malawi are smallholder farmers. Their farms average about 1.5 hectares and produce as many as twenty crops in the course of a year, using various sorts of rotation and integration. These farms differ widely from place to place, depending upon a wide variety of social, economic and environmental circumstances. This group of farmers has, in general, failed to benefit from the "green revolution" agricultural innovations developed by researchers who largely focused on commercial monocropping systems.

INTEGRATED RESOURCE MANAGEMENT

Integrated resource management (IRM) is the use of waste products from one agricultural enterprise to fuel another, thus creating economic, environmental and productivity synergisms which improve overall farm efficiency and profitability. IRM systems have been shown to operate effectively in test cases in Malawi

(Brummett & Noble, 1995a), Ghana (Prein, 1995) and the Philippines (Lightfoot & Pullin, 1995). In addition to immediate improvements in farm function, longer-term ecological benefits such as reduced soil erosion and increased tree cover are predicted (Lightfoot & Pullin, 1995).

Creating the maximum impact of these systems, however, requires that farmers have an holistic vision of their farm which permits the efficient management of resources. Unfortunately, many small-scale farmers rely more on tradition than on any systematic approach to farming in making resource allocation decisions. Brummett and Noble (1995a) found, for example, that many farm enterprises are carried from year to year, despite continually losing money or earning only negligible amounts. Both the complexity of small farms and the conservatism of smallholding farmers serve to dramatically increase the number of factors which must be weighed in decision-making. The simple reliance upon production and net return calculations which most agriculture researchers use to promote their technologies often falls short of convincing the smallholder. The tools and methods of IRM (cf. Lightfoot & Noble, 1993; Brummet & Noble, 1995b; Prein, 1995) are designed to take the complex reality of small-scale farms into consideration when designing improved farming systems.

The concept of IRM goes beyond simple integration and technology transfer. The farmers' adoption of better techniques may increase farm profitability and food production in the short term. However, unless these farmers also adopt a spirit of innovation and openness to improved farming methods, the gain will be lost over time as the population continues to grow. Smallholding farms must evolve if they are to continuously improve performance and become economically and ecologically viable and sustainable.

Farmer-Scientist Research Partnerships

Poorly educated small-scale farmers, operating within the perspective of a rural ecosystem which incorporates a large number of unquantifiable social and environmental factors, often have difficulty explaining their situation clearly to researchers, who too often use their advanced schooling to focus on the details of maximizing the output from their special crop. A first step in designing new, more appropriate technologies and giving farmers the mental tools to adopt a more progressive approach is to develop mechanisms for the interactive exchange of information and ideas between farmers and researchers.

Resource flow diagramming

In designing research projects to develop more appropriate technologies for smallholders, ICLARM uses resource flow diagramming as a basis for communication between farmers and researchers. The diagramming exercise generally begins with an interview, during which the proposed activities are presented to farmers and their feedback is solicited. Over the course of the discussions, farmers who are truly interested in participating in collaborative research can be separated from those who wish only to be part of a development activity (Harrison, 1995). Following these initial discussions, a walking tour of the farm can enable researchers to roughly characterize the farm for inclusion or exclusion from particular studies. For example, Brummett and Chikafumbwa (1995) interviewed nine farmers and surveyed their lands prior to selecting six for participation in a rain-fed aquaculture study. The three excluded farms had access to permanent water supplies, making them unsuitable for that study.

Once farmers who are both interested and fit the experimental design are identified, resource flow diagramming is used to characterize the farms in terms of their resource base. It also provides a means for giving the farmers a systematic perspective on their farming system, which they may have never had before, and which might arguably be a prerequisite for efficient farmer experimentation (Lightfoot & Noble, 1993).

Once the main features of the farming system are noted, farmers are asked to describe how the various enterprises depicted relate to each other. This is done by connecting the various enterprises with arrows to show the direction of resource movement. If, for example, stovers from a maize crop are used to mulch a vegetable garden, a flow arrow is drawn on the map to connect the maize and vegetable plots, indicating the direction of the resource movement (in this case towards the vegetables). Materials which are eaten by the families would be connected to the house with an arrow. Materials which are sold would be connected to an external market. Cash (generally payment for labour, or money earned on produce sales), fertilizers and animal feeds are included as resource flows which connect the farm to the local economy. For greater detail, values can be attached to flows, to give them a quantitative dimension (Lightfoot & Noble, 1993).

Figure 19.1 is such a map drawn by Mrs. D. Ng'ombe of Zomba District in Southern Malawi. This particular map shows only the various farm enterprise

Figure 19.1 A resource flow diagram drawn by Mrs. D. Ng'ombe of Zomba District, Southern Malawi, showing her farm management strategy for the 1992-1993 growing season.

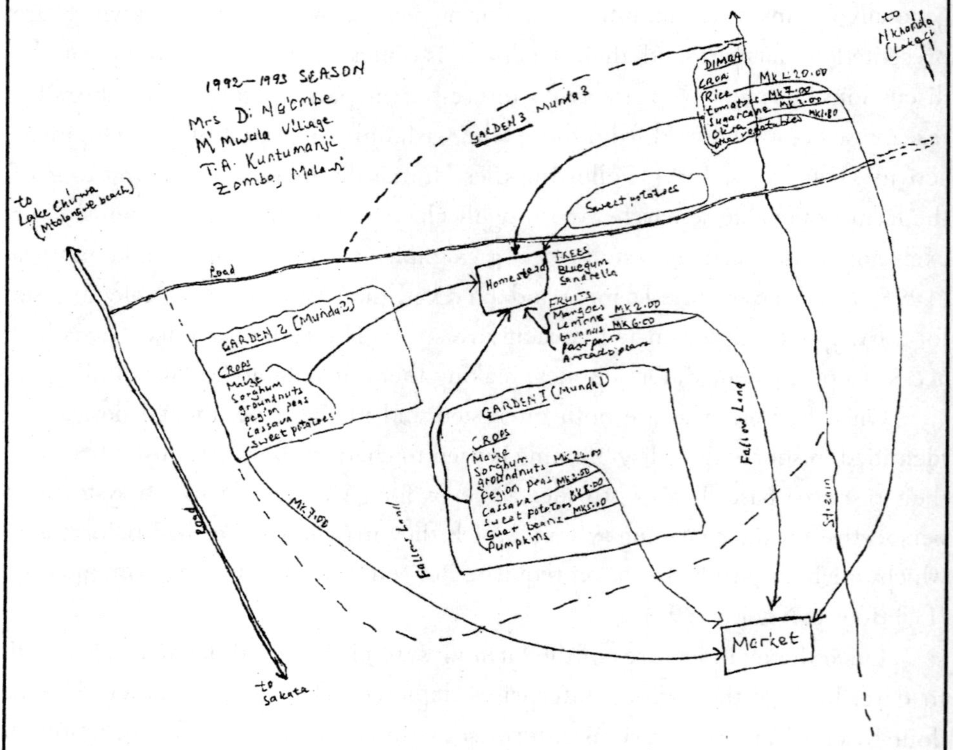

Garden 1 crop:	Garden 2 crops:	Trees	Fruits
Maize	Maize	Bluegum	Mangoes
Sorghum	Sorghum	Sandralle	Lemons
Groundnuts	Groundnuts		Bananas
Pegion peas	Pegion peas		Pawpaws
Cassava	Cassava		Avocada pears
Sweet potatoes	Sweet potatoes		
Sugar cane			
Pumpkin			

systems and the movement of resources around the farm and into the surrounding economy. Depending upon the end purpose, details of soil type, slope and water resources can be easily added. Such mapping provides the researcher with a detailed picture of the diversity and distribution of land, soil and water resources from the farmer's perspective. Farmers gain a perspective, often for the first time, of their own farm's relationship to the surrounding rural community and its agro-ecological environment.

Farmer-led experimentation

Having established a map showing the enterprises and resource flows on the farm, the farmer is requested to imagine a scenario where a new or modified enterprise is incorporated into the farming system. In the case of ICLARM, researchers were trying to introduce the idea of IRM through the incorporation of a fish pond into the existing farming system. Once the new idea has been presented in general terms, the resource flow diagram is re-drawn to show the theoretical relationships between the new activity and existing ones. Figure 19.2 shows such a diagram for Mrs. Ng'ombe. The redrawing of the map gives the researchers the opportunity to discuss the specific details of IRM. It is interesting to note that only the fishpond has links with other resource systems and enterprises which do not pass through the household or an external market. Fish ponds have a particularly high capacity for using and transforming agricultural wastes without creating pest or human health problems (Lightfoot, Bimbao, Dalsgaard & Pullin, 1993).

The theoretical farming system model created during the re-drawing session is used by farmers and researchers as a guide for conducting applied experiments. In the case of integrated agriculture-aquaculture, the farmer constructs the pond in a site selected in consultation with the researcher (to make sure that it will, for example, fill with water). Other than general advice, no inputs to pond construction are provided by the researcher. Once the pond is constructed and full of water, the researcher must provide the fingerlings for fish stocking, to ensure that the quantity and quality of seed are known and controlled. The farmer then uses the resource flow diagram to manage the farm. The farmer records the amounts of materials which flow along the different pathways and notes any deviations from the design.

To control (in the scientific sense) the experiments conducted on farm, replicated simulations are established on the experiment station. Every week,

Figure 19.2 Resource flow diagram for 1992-93 by Mrs. D. Ng'ombe, Zomba

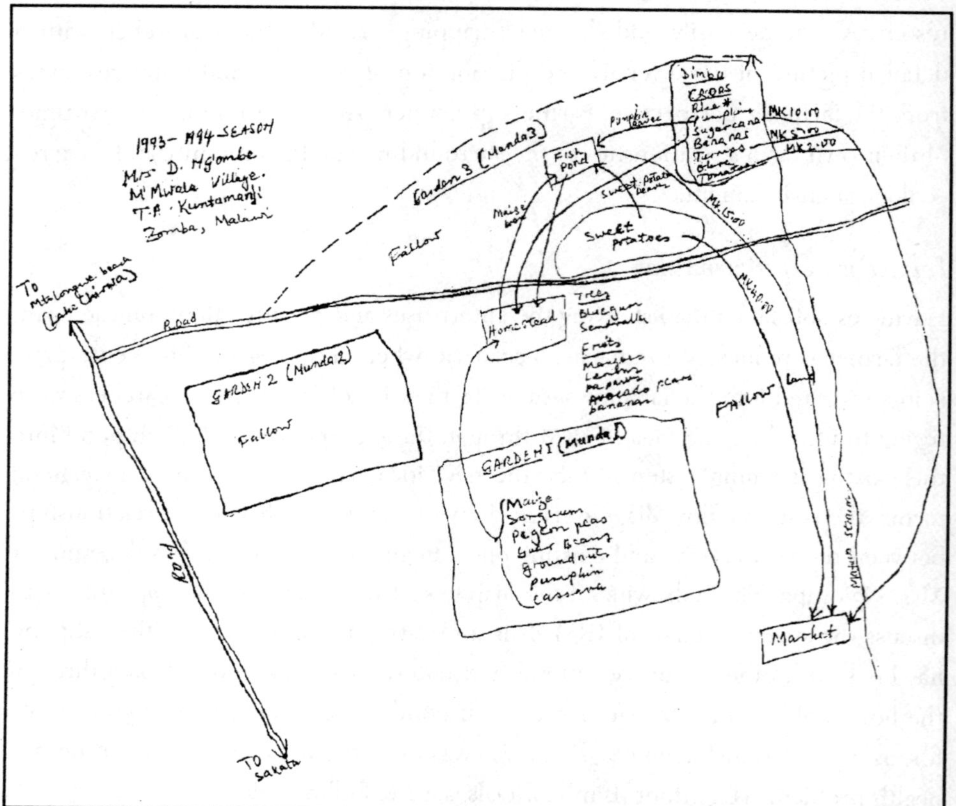

samples of on-farm inputs are collected, dried and analyzed. This data is then loaded into an ICLARM-developed spreadsheet which reduces to dry matter, organic matter, nitrogen and phosphorus terms the inputs used by farmers. Through a comparison of this data with observations made on-farm, inputs and management practices used by farmers can be mimicked on the experiment station. This system provides the statistical comparisons needed to make general statements about farm management strategies. It also gives the researchers an opportunity to experience, at a personal level, the problems faced by farmers. This helps provide insight to potential new technologies and, perhaps more importantly, provides the shared experience central to a more positive mutual understanding between scientist-researcher and farmer-researcher. Such a relationship is key to a more realistic and fruitful research and development methodology.

Technology Development

As Figure 19.3 illustrates, the conduct of farmer-led experimentation presents the opportunity to perform on-station research and development which is inherently appropriate to the farmer's situation. In classical approaches, potential management strategies are compared to some sort of best-case scenario. Researchers tend to choose systems for study which push the productivity of the fish pond, without regard for whether or not farmers really have access to the inputs necessary to use the technologies developed on their farms.

In the Farmer-Scientist Research Partnership approach, the data gathered during weekly visits, as mentioned above, is used to establish scientific controls of on-farm studies conducted by farmers. These controls are also used by scientists on the experiment station to conduct studies of improved systems. For example, if

Figure 19.3 Farmer-Scientist research partnership model

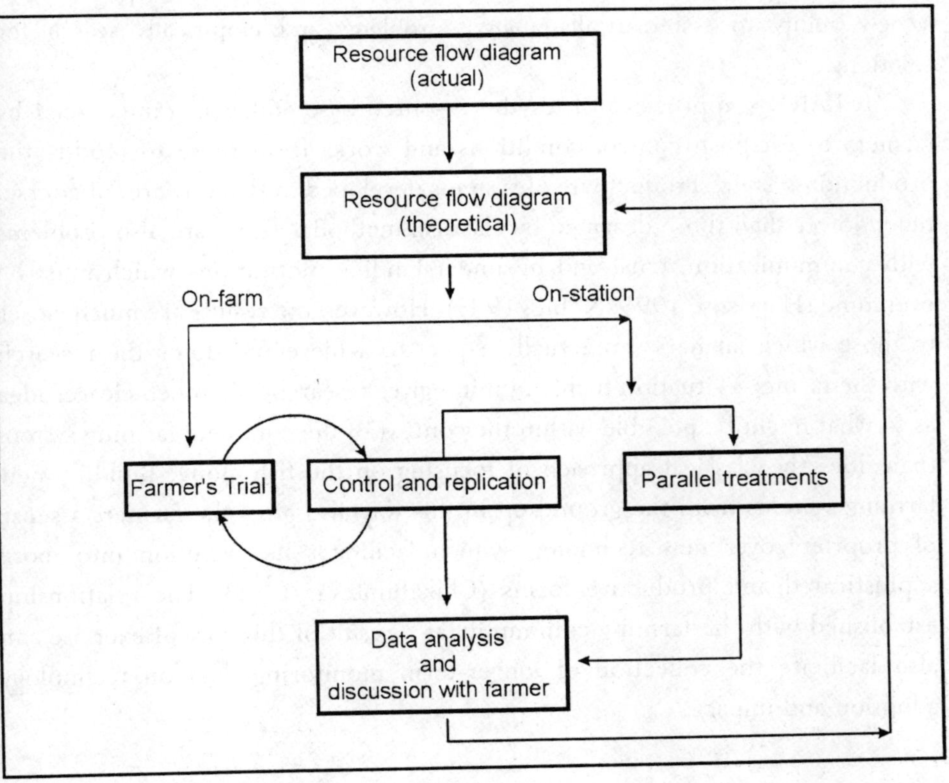

a farmer elects to test the use of chopped maize stover as a pond input, the researchers establish a maize stover control which simulates the on-farm treatments. Then the researchers design other treatments which are also controlled by the same ponds, which control the on-farm study. Such treatments might include the use of different presentation strategies for the maize stover, such as grinding, composting, etc.

When the growing season is over, all ponds both on-farm and on-station are harvested and the results analyzed and compared. The outcome is presented to the farmer for discussion. Sometimes, ingenious methods are required to clearly demonstrate to the farmers what has happened and care must be taken to avoid confusion (Hopkins, 1988). Group discussions among participating farmers are often useful (Lightfoot & Noble, 1993).

Based on the results of both the farmer's and researcher's studies, farmers are requested to re-draw their resource flow diagrams, to show how the system will be managed in the following year. Depending on the objectives of the research program, this cycle can be repeated and may, over time, help to develop in farmers a new ability to systematically analyze problems and empirically search for solutions.

ICLARM's approach utilizes the resource base and constraints faced by farmers to establish control conditions and works from there to modify the production system. Productivity of systems developed in this way are, of course, much lower than those designed by classical methods. There are also problems with communication, trust and misunderstanding motivations which must be overcome (Harrison, 1995; Noble, 1995). However, the results are much closer to those which farmers can actually expect to achieve and doing the research with the farmer's situation firmly in mind gives researchers a much clearer idea as to what might be possible within the context of the complete farming system than does the classical approach of focusing on the fish alone. Building new farming systems from the ground up in this way also gives the farmers a sense of propriety over new technology which facilitates its evolution into more sophisticated and productive forms (Chikafumbwa, 1995). The relationship established with the farming community as a result of this sort of exercise can also facilitate the collection of longer-term monitoring data on technology adoption and impact.

Conclusions

ICLARM uses the farmer-scientist research method to study smallholding farming systems and devise innovative approaches to problems that constrain agricultural productivity. However, the farmers themselves use the outputs of the method for more than just one season of research. In the words of one farmer with whom ICLARM has worked for several seasons: "Participating in the research project has been helpful because I have learned how to keep records of activities and now know how my farm enterprises are being run in terms of labour cost and input costs. I now realize which enterprises are bringing more money to the family and which ones are not helpful."

Although not specifically designed to transfer technology, the adoption rate of the IRM-based aquaculture innovations among farmers who participated in farmer-scientist research partnerships has varied between 65% and 100%. More remarkably, farmers involved in the development and testing of the methodology have continued to expand the aquaculture components of their farms over the past three years, despite droughts that have dramatically reduced the productivity of their fish ponds.

Whether the hoped-for spirit of evolution has been instilled in these farmers remains to be seen, but preliminary indications are good. Lightfoot and Noble (1993) and Chikafumbwa (1995) noted that farmers with whom researchers had interacted had adopted and modified the technologies they had helped to develop – and had each spread these technologies to at least four other farmers.

Endnotes

1. This work was conducted with financial support from the German Bundesministerium fur Wirtschaftliche Zusammenarbeit (BMZ) and the German Gesellschaft für Technische Zusammenarbeit (GTZ).

References

Brummett, R.E. 1995: The context of smallholding integrated aquaculture in Malawi. In R.E. Brummett (ed.), *Aquaculture Policy Options for Integrated Resource Management in SubSaharan Africa*. ICLARM Conf. Proc. 46. Manila: International Centre for Living Aquatic Resources Management.

Brummett, R.E. and Chikafumbwa, F.J.K. 1995: *Management of rainfed aquaculture on Malawian smallholdings.* Unpublished paper presented at the FACON Conference on Sustainable Aquaculture, 11-14 June, Honolulu, Hawaii. Pacific Congress on Marine Science and Technology.

Brummett, R.E. and Noble, R.P. 1995a: *Aquaculture for African smallholders*. ICLARM Tech. Rep. 46. Manila: International Centre for Living Aquatic Resources Management.

Brummett, R.E. and Noble, R.P. 1995b: Farmer-scientist research partnerships and smallholder integrated aquaculture in Malawi. In J-J. Symoens and J-C. Micha (eds.), *The management of integrated freshwater agro-piscicultural ecosystems in tropical areas.* Wageningen and Brussels: Technical Centre for Agricultural and Rural Cooperation and the Royal Academy of Overseas Sciences.

Chikafumbwa, F.J.K. 1995: Farmer participation in technology development and transfer in Malawi. In R.E. Brummett (ed.), *Aquaculture Policy Options for Integrated Resource Management in SubSaharan Africa.* ICLARM Conf. Proc. 46. Manila: International Centre for Living Aquatic Resources Management.

Harrison, E. 1995: Digging fishponds: perspectives on motivation. In R.E. Brummett (ed.), *Aquaculture Policy Options for Integrated Resource Management in SubSaharan Africa.* ICLARM Conf. Proc. 46. Manila: International Centre for Living Aquatic Resources Management.

Hopkins, K.D. 1988: Reporting fishpond yield to farmers. *Aquabyte* 1 (2), 6.

Lightfoot, C. and Noble, R. 1993: A participatory experiment in sustainable agriculture. *J. Farming Systems Rsch. Ext.* 4 (l), 11-34.

Lightfoot, C. and Pullin, R.S.V. 1995: Why an integrated resource management approach? In R.E. Brummett (ed.), *Aquaculture Policy Options for Integrated Resource Management in SubSaharan Africa.* ICLARM Conf. Proc. 46. Manila: International Centre for Living Aquatic Resources Management.

Lightfoot, C., Bimbao, M.A., Dalsgaard, J.P.T. and Pullin, R.S.V. 1993: Aquaculture and sustainability through integrated resources management. *Outlook on Agriculture* 22 (3), 143-150.

Noble, R.P. 1995: Research challenges in integrated resource management (IRM) in rural Africa. In R.E. Brummett (ed.), *Aquaculture Policy Options for Integrated Resource Management in SubSaharan Africa.* ICLARM Conf. Proc. 46. Manila: International Centre for Living Aquatic Resources Management.

Prein, M. 1995: Farmer participatory development of integrated agriculture aquaculture systems for natural resource management in Ghana. In R.E. Brummett (ed.), *Aquaculture Policy Options for Integrated Resource Management in SubSaharan Africa.* ICLARM Conf. Proc. 46. Manila: International Centre for Living Aquatic Resources Management.

CHAPTER 20

CONTROLLED MATING OF *OREOCHROMIS SHIRANUS* AND *TILAPIA RENDALLI* FOR STOCK IMPROVEMENT IN MALAWI.

by Alfred O. Maluwa and Aggrey D.J. Ambali

INTRODUCTION

Aquaculture is playing an important role in supplying fresh fish to the rural communities of Malawi, especially those in the areas most remote from the lakes. Due to a population increase of 3.2% per annum (National Statistical Office, 1977) and stagnation of fish catches from the country's natural waters at 70,000 metric tonnes per annum, per capita fish consumption in Malawi declined from 14kg in 1970 to 7kg in 1992. The result is that fish has become very scarce and expensive for most rural households.

Apart from its natural waters, Malawi has great potential for the development of aquaculture at both small-scale and commercial levels. It is estimated that more than 11,650 km^2 of land in Malawi is under or has potential for aquaculture; this is about 15% of the total available land (Brooks, 1992). Using Geographical Information System techniques, Kapetsky (1994) calculated that for subsistence level fish farming, 55,905 km^2 and 8,994 km^2 of land in Malawi is suitable for farming one crop per year with good growth and two crops per year with fair growth, respectively. Following a 1993 FAO study, it was concluded that lack of suitable areas is not a constraint on the further development of aquaculture in Malawi.

The history of aquaculture in Malawi dates back to the 1930s, however most the expansion to rural households occurred between 1970 and 1980, when aquaculture projects were externally funded. There are now over 3,000 farmers owning over 4,000 fish ponds. The species introduced are tilapias indigenous to Malawian waters; *Oreochromis shiranus* and *Tilapia rendalli*. The major limitation to increased fish production has been the slow growth rates of these tilapias.

Tilapias breed naturally in ponds and their seed production on both government and private farms is routine, hence, genetic improvement of the quality of seed from these stocks is possible. To date, breeding methods based on quantitative genetics have contributed little to improvement of tilapia growth rates because most applied research has focused on producing monosex tilapia populations to check the precocious reproduction behavior of the species (Guerrero, 1975; Wohlfarth & Hulata, 1983) rather than on selection for growth characteristics.

The Study

The research work reported here is a component of the main tilapia breeding/genetics programme at the National Aquaculture Centre, Domasi, Malawi. The tilapia programme is trying to contribute to sustainable agriculture/aquaculture systems by improving tilapia growth rates through selection. The study reported here served to form a base population on which further genetics studies will be done.

A major concern in any breeding program is the proper choice of a selection procedure. In genetic studies, it is difficult to differentiate between potential environmental sources of variation and variance based on family characteristics. With fish breeding, the identification of individuals at a small size is very difficult. They must, therefore, be raised as families. When families are grown separately, family identity is maintained and selection for desirable characteristics can be done within each family. Working with families also makes control for environmental variables easier, as the time of spawning, feeding rates, water flows do not have to be strictly controlled between families (Abella, Palada & Newkirk, 1989). In tilapias, breeding is asynchronous. This introduces another source of variation. There will be a size-at-age bias tending to associate larger with older individuals. But within a family, selection can also control for this effect because all individuals within the family are the same age. The objective of the current study was to select for growth characteristics in both domesticated and wild *O. shiranus* and *T. rendalli.* For growth characteristics, the criterion used was the deviation of the individual from the mean value of the family to which it belongs. Those individuals that exceed their family means by the greatest amount are regarded as the most desirable (Abella *et al.*, 1989).

SPAWNING PROCEDURE

Controlled mating and rearing of first generation offspring (F1) was the first step in the selection exercise. Broodstock for domesticated *Oreochromis shiranus(D)* and *Tilapia rendalli(D)* were obtained from ponds at the National Aquaculture Centre. These fish have been domesticated for 3 years. Broodstock for wild *Tilapia rendalli (W)* were collected directly from Lake Chiuta. Broodstock for wild *Oreochromis shiranus(W)* were collected from a lagoon on the shores of Lake Malawi, in the Central region (Nkhota kota). The spawning procedure followed a single pair mating system. Twenty-four tanks of $1 m^3$ were used, 12 tanks for domesticated and 12 for the wild stocks. For each strain, wild or domesticated, both males and females were allocated numbers from 1 to 12. In the first set-up, male 1 was allocated female 1, male 2 to female 2, and so on, up to 12. When the fish spawned, the male and fry were removed from the tank and this male was allocated to another available female, depending upon which other tank had spawned. Care was taken to ensure that a given male spawned with a given female only once. Fry were removed and labeled according to the parental code, e.g. (1x1) stands for fry bred from male number 1 and female number 1. This was done in both species and strains, viz: *O.shiranus(D), O. shiranus(W)* and *T rendalli(D)* and *T. rendalli(W)*, and each species and strain had its own numbers from 1 to 12 for both males and females. As a result, the fry from *T. rendalli* and *O. shiranus* had a combination of numbers from 1 to 12 with a subscript D for domesticated and W for wild strains. The fry were raised as separate families in hapas of $1 m^3$. After 2 weeks, when the fry attained 2 g, the number of juveniles was set at 300 per $1 m^3$ hapa and were grown for 3 months (90 days).

SIZE GRADING (COLLIMATION) AND GROWTH TESTING

First selection

After 90 days, the first size grading was undertaken. The size grading (collimation) technique aims at selecting fish of the same size. To select individuals of the same size, the fish total length (TL) data for 4,800 individuals from each strain was measured to + 0.1 cm. The data were entered into Excel spreadsheets of Windows 95, Microsoft Office software, installed in a Pentium Personal Computer. The TL means (μ) and standard deviations (δ) were calculated using the formulae in the spreadsheets. With 4,800 records, it was possible for the length frequency distributions to be standardized into a normal distribution curve (Zar, 1986). The standard deviation on both sides of the mean was used as a criterion for choosing uniformly sized fingerlings around the mean TL values. With this selection strategy, any differences between family means which may be due to the environment were irrelevant (Falconer, 1982). And because the selected fish were of the same size despite their age, the size-at-age bias was removed (Doyle and Talbot, 1986b). The selected individuals were tagged to maintain their family identity.

Second selection

The selected fingerlings were raised for another 90 days in a mass grow-out pond of 200m^3. Selection for growth was performed after this second grow-out period. Total length data for a sample of 500 individuals from each strain was used to determine the frequency distribution curve. Those individuals whose TL fell in the top 5% of the frequency distribution were selected as a base population for future breeding work at the centre. The identification of this high-growth set constituted the final stage of the study.

FEEDING

The broodstock in tanks were fed a formulated feed whose composition is shown in Table 20.1. The feed had about 20% protein.

Table 20.1 Composition by ingredients of brookstock feed in tanks

Ingredient	(%)
Meat and bone meal	25.0
Soy bean meal	25.0
Wheat flour	25.0
Fish meal	10.0
Maize bran	13.5
Vitamin mixture	1.0
Mineral prix 10.0	0.5

This feed was given in powder form at 3% body weight per day and given 3 times (morning, midday and afternoon) every day. The composition of formulated feed given to fry in tanks and hapas is shown in Table 20.2. The protein content of the feed was about 30%.

Table 20.2 Composition of fry feed

Ingredient	(%)
Fish meal	50.0
Meat and bone meal	12.5
Wheat flour	25.0
Milk powder	10.0
Vitamin mixture	2.0
Mineral prix 50.0	0.5

POND FERTILIZATION

The ponds in which the hapas were inverted were fertilized with chicken manure at the rate of 50 kg dry matter per hectare per day. The fry feed was given as a supplement. The mass grow-out ponds were fertilized with chicken manure at 50 kg dry matter per hectare per day, and maize bran was fed as a supplement at 3% body weight per day.

WATER QUALITY

In tanks for both broodstock and fry, the following water quality parameters were monitored every week: Ammonia, Nitrate and pH. Dissolved oxygen (DO) and temperature were monitored every day. In ponds, DO and temperature were monitored every day and pH and phosphorus every month.

RESULTS

Tilapia rendalli (Domesticated)

The single pair mating schedule which produced *Tilapia rendalli (D)* fry is shown in Table 20.3. There were 36 F1 families in total. The frequency of spawning per female ranged from 2 to 5. Female number 4 and 10 bred only once, while 1 and 9 bred twice. Females 3, 8, and 12 bred three times and females 2, 5, 6, and 7 had the second highest breeding frequency of 4 times. The most prolific spawner was female number 11, which spawned 5 times.

Table 20.3 Results of single pair mating of *Tilapia rendalli (D)* producing F1 offspring

Female	Male											
	1	2	3	4	5	6	7	8	9	10	11	12
1	1xl						1x7					
2		2x2	2x3								2x11	2x12
3		3x2	3x3							3x10		
4				4x4								
5				5x4	5x5			5x8				5x12
6				6x4	6x5	6x6			6x9			
7				7x4	7x5	7x6	7x7					
8			8x3					8x8	8x9			
9								9x8	9x9			
10										10x10		
11						11x6	11x7		11x9		11x11	11x12
12										12x10	12x11	12x12

The length-frequency distribution at size collimation for this strain is shown in Figure 20.1. The mean total length was 7.83cm, and the standard deviation was 1.19cm. The chosen individuals were one half a standard deviation on both sides of the mean or 60cm on each side of the mean (1.19/2 = 0.60). As a result all fish with total length between 7.2 and 8.4 cm were chosen, tagged and grown in mass grow-out ponds (7.83 - 0.60 = 7.2cm and 7.83 + 0.60 = 8.4cm). The selected individuals were 30% of the total population.

Figure 20.1 Length - Frequency distribution of *Tilapia rendalli* (Domesticated) during size collimation. (μ = 7.83cm, δ = 1.19cm, n = 4,800)

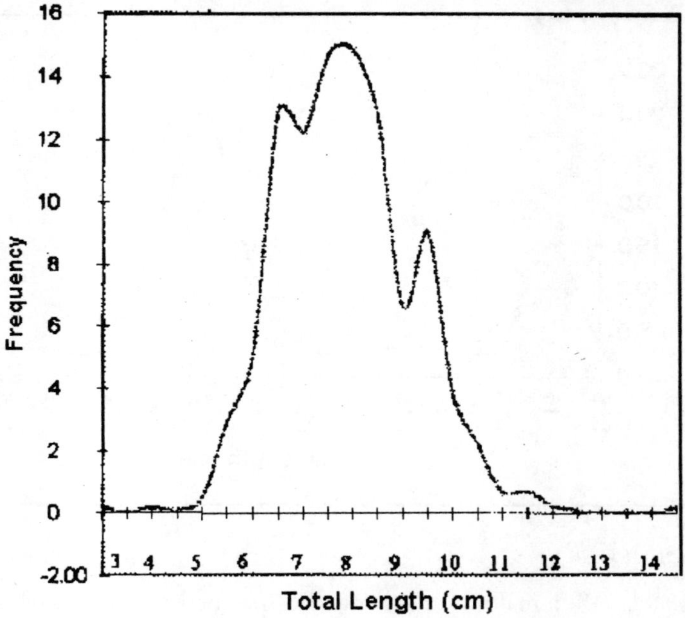

The size distribution of 500 fish from the second grow-out period in the mass grow-out 200m^2 pond is shown in Figure 20.2. The mean of the distribution was 18.8cm and the standard deviation was 1.61. Upon standardization, it followed that the longest individual was 1.7 standard deviations from the mean (21.6 - 18.8)/1.61 = 1.7. One standard deviation from the mean were individuals above 20.4 cm and those fish with 20.4 cm total length were selected as parents of the next generation. These were 27.2% of the total population and their TL fell into the top 5% of the distribution curve.

Figure 20.2 Length-frequency distribution of *Tilapia rendalli* (Domesticated) after 90 days of growth testing in mass grow-out ponds (μ = 18.8 cm, δ = 1.61 and n = 500)

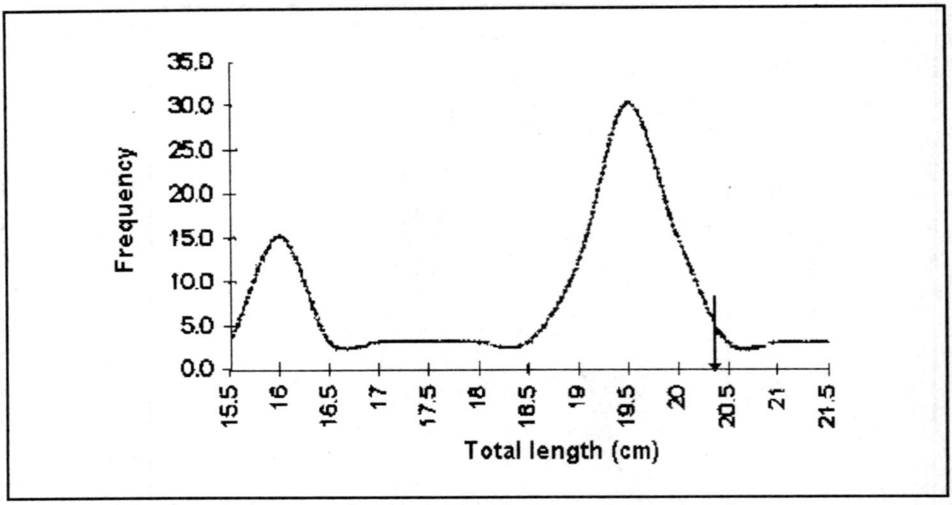

Tilapia rendalli (Wild)[1]

For *Tilapia rendalli (W)* females number 5 and 6 spawned only once, while 2, 7 and 11 spawned twice, 9 and 10 three times, 3, 4 and 12 five times and female number 8 spawned a record 7 times. A total of 40 families were raised. At size collimation the mean length of the 4,800 sampled was 6.87cm and the standard deviation was. 0.72 cm. Individuals whose lengths fell between 6.15 and 7.59 cm were chosen, tagged and then reared in a mass grow-out pond. The fingerlings from this wild strain of *T. rendalli* had less size variation than the domesticated strain (0.72 cm versus 1.2 cm). As such, one standard deviation was used for the wild strain, instead of the one half standard deviation that was used with the domesticated strain. The harvesting mean total length for *T. rendalli(W)* in the mass grow-out pond was 19.89 + 1.13. After the second grow-out period individuals with length greater than 20.5 cm fell in the top 5% of the standard normal distribution curve and these have been selected as a base population for future breeding programmes on the station.

Oreochromis shiranus (Domesticated)

With *O. shiranus(D)* females 6 and 7 spawned twice and females 10, 11 and 12 never spawned. The rest of the fish took a long time to spawn in the breeding tanks. In general, breeding of *O. shiranus(D)* in a 1 m^3 concrete tank proved very difficult. However, there are reports that the fish has been bred quite easily in hapas on the station. These breeding trials should be tried out in hapas instead of concrete ponds.

Eleven families were raised. At size collimation the mean total length was 6.5 cm and the standard deviation was 1.9cm. All juveniles whose lengths fell between 5. 1 and 7.9 cm were chosen and tagged in order to maintain their family identity and reared in mass grow-out ponds. At harvest, the mean total length of the 500 sampled was 16.3 cm and the standard deviation was 2.1 cm. All the fish with a total length longer than 18.3 cm fell in the top 5% of the frequency distribution curve. These were chosen as broodstock for the next generation.

Oreochromis shiranus (wild)

For *O. shiranus(W)* a total of 13 F_1 families have been raised from females 2, 6, 7 and 9 which spawned twice. There was no spawn from females number 11 and 12, while the rest spawned only once. The mean total length after 90 days of rearing in hapas was 5.9 cm and the standard deviation was 1.7 cm. All the juveniles whose total length ranged between 4.2 and 7.6 cm were tagged according to their families and stocked in a mass grow-out pond. At harvest, after 90 days all the fish whose total length was longer than 18.7cm fell in the top 5% of the frequency distribution curve and these were chosen as broodstock for the next generation.

Water Quality

In tanks, ammonia readings ranged from 0.004 to 0.007 mg/1, nitrate 0.06 to 0.93 mg/1, pH 7.2 to 9.0 and for dissolved oxygen, the lowest reading was 2.63 mg/1 and the highest was 7.2 mg/1. The temperature range was 19.8 to 27.5C. Temperatures lower than 19C limited the breeding of *O. shiranus* but not *T. rendalli*. These low temperatures were recorded during the cold months, July and August. Thereafter, the temperatures were within the acceptable limits for *O. shiranus* breeding. In ponds and hapas, the temperature remained above 24C. Dissolved oxygen readings remained above 3 mg/1 and phosphorus ranged from 0.9 to 2.5 mg/l. All the water quality measured were within the optimum range for tilapia breeding and growth (Balarin & Haller, 1982).

Discussion

Breeding of *Tilapia rendalli* in tanks has been easier than breeding of *Oreochromis shiranus*. The *Tilapia rendalli* produced 36 and 40 families for domesticated and wild strains respectively, compared to 11 and 13 families produced from domesticated and wild strains of *O. shiranus*. Maluwa (1990) reported breeding of *O. shiranus* every 14 days in hapas at the National Aquaculture Centre, provided the temperatures were higher than 19C. Ambali (1990) reported the interspawning period for *O. shiranus* in hapas at temperatures higher than 19C to be between 10 and 14 days when broodstock was conditioned. In the current study, the temperatures in the tanks were in the desirable range, but breeding was not as frequent as reported for the hapas. It will therefore be worth trying the single pair mating in hapas for more families of *O. shiranus* juveniles. Between the two *T. rendalli* strains it has been noted that there was more size variation in the wild than in the domesticated strain. This is evident in both collimated and uncollimated stocks.

This research project has been instrumental in the formation of a base population on which further genetics studies will be done. The top 5% from each distribution curve were selected to form the parents for future studies. Future research work will involve breeding the best fish obtained in this research project and evaluating the juvenile growth rates in ponds, using the unselected fish as a control.

Endnotes

1 Editor's note: The original report contained mating schedule tables, and graphs for length- frequency distribution at size collimation and for length-frequency after 90 days in grow out ponds for each of the four strains being studied. Only the table and graphs for *Tilapia rendalli* (domesticated) have been produced here. The three other sets were similar in presentation. All data on means and standard deviations are contained in the text.

References

Abella, T.A., Palada, M.S. and Newkirk, G.F. 1989: Within Family Selection for Growth Rate with Rotational Mating in *Oreochromis niloticus*. In R. Hirano and I. Hanyu (eds.), Proceedings of the *2nd Asian Fisheries Forum*, pp. 515-522. Asian Fisheries Society.

Ambali, A.J.D. 1990: Effect of Broodstock Conditioning on the Frequency of Spawning of *Oreochromis shiranus* in hapas. Unpublished. Zomba: University of Malawi, Chancellor College.

Balarin, J.D. and Haller, R.D. 1982: Intensive Culture of Tilapia in tanks, raceways and cages. In J.F. Muir and R.J. Roberts (eds.), *Recent Advances in Aquaculture.* London: Croom Helm Ltd.

Brooks, A.C. 1992: *Viability of Commercial Fish Farming in Malawi.* Fisheries Department Project Report. Lilongwe: Ministry of Natural Resources.

Doyle, R.W. and Talbot, A.J. 1986b: Artificial Selection on Growth and Correlated Selection on Competitive Behavior in Fish. *Can. J. Fish. Aquat.* Sci. 43, 1059-1064.

Doyle, R.W. and Talbot, A.J. 1989: Repeatability of Relative Size Specific Growth in tilapia. In R.S.V. Pullin, T. Bhukaswan, K. Tonguthai and J.L. MacLean (eds.), *The Second International Symposium on Tilapia in Aquaculture*, pp. 451-456. Bangkok and Manda: Department of Fisheries and ICLARM.

Facolner, D.S. 1982: *Introduction to Quantitative Genetics.* Second Edition. Department of Genetics and Agricultural Research Council Unit of Animal Genetics, University of Edinburgh. London and New York: Longman.

Guerrero, R.D. 1975: Use of Androgens for the Production of all-male *Tilapia aurea*.(Stendachner). Trans. Am. Fish. Soc. 103, 342-348.

Kapetsky, J.M. 1994: The Potential for Warm Water Fish Farming in the SADC Countries. *FAO Aquaculture Newsletter*, April, 1994, No. 4, 2-6. Rome: FAO, Inland Water Resources and Aquaculture Service.

Maluwa, A.0. 1990: *Reproductive Biology and Fry Production of* Oreochromis shiranus. M.Sc. Thesis. Zomba: University of Malawi.

National Statistical Office. 1977: *Malawi Human and Population Census.* Zomba: Government Printer.

Wohlfarth, G.W. and Hulata, G. 1983: *Applied Genetics of tilapias.* Second Edition. ICLARM Studies and Reviews 6. Manila: ICLARM.

Zar, J.H. 1986: *Biostatistical Analysis.* Second Edition. Englewood Cliffs: Prentice-Hall Inc.

CHAPTER 21

POSTHARVEST SYSTEMS IN UGANDA: THE PROCESSING AND STORING OF HORTICULTURAL COMMODITIES

by Archileo N. Kaaya

INTRODUCTION

Agriculture is the mainstay of Uganda's economy, with over 90% of its economically active population either greatly or entirely dependent on this industry for their income and livelihood. About 74% of the agricultural gross product is food crops, of which about one-third is marketed, the rest being produced for subsistence. Fruit and vegetable crops constitute the largest segment of food crop production. The quality of fruits and vegetables, including appearance, texture, flavour, and nutritive value, are susceptible to postharvest deterioration.

In Uganda, postharvest food losses are conservatively estimated to be 30-40% of total production. Any approach to ecological agriculture should take into account the problem of postharvest losses and pay attention to the need for effective processing, preservation and storage. Postharvest changes in fresh produce cannot be stopped, but they can be slowed. In developed countries, refrigerated storage is used as well as, more recently, more sophisticated techniques such as modified atmosphere and hypobaric storage, which provide an optimal or near-optimal environment for the produce. Large-scale operations may benefit from investing in such costly handling machinery and high-technology postharvest treatments, but these options are not available to small-scale producers. Simple, low-cost technologies are more appropriate for small-volume commercial operations and farmers involved in direct processing, storage and marketing.

The food processing industry is among the fastest growing sub-sectors of manufacturing in Uganda. Currently, it contributes approximately 32.7% to the index of industrial production. It contributes to food security and to foreign exchange saving through efficient import substitution. Its operations include grain milling, baking, brewing, bottling of soft drinks, distilling, sugar production, edible oil production, meat and fish processing, dairy production, animal feed production, and coffee and tea processing. Horticultural crop processing is still in its infancy.

Uganda has only three large-scale fruit and vegetable processing firms: RECO, in Kasese district, produces jams, marmalades, papain and fruit concentrates; Elgonia Industries, in Tororo district, produces passion and orange squash, tomato and chilli sauces; Megatrends, in Kampala, produces tomato and chilli sauce. But there are many cottage-scale enterprises processing and preserving fruits and vegetables. Research on several areas of processing these commodities is currently underway at Kawanda Agricultural Research Institute of the National Agricultural Research Organisation of Uganda (NARO) and at the Department of Food Science and Technology and the Department of Chemistry of Makerere University. The primary objective is to improve existing small-scale technologies and increase crop utilization. This paper gives an overview of some methods suitable for the small-scale producer for processing and storing horticultural commodities. The following groups of horticultural commodities are considered:

- **Fruits:** banana, pineapple, passion fruit, tomato, mango, orange
- **Root crops:** cassava, sweet potato
- **Leafy vegetables:** cabbage, cowpea leaves, amaranth, various solanaceous plants
- **Spices:** chilli, ginger, onion, garlic

The postharvest handling of grains and oil seeds, as well as other products such as mushrooms, are not included in this paper.

Fruit Processing

The supply of fruits grown in Uganda is not constant throughout the year. Some, like mangoes, may be unavailable during certain seasons, while others, like bananas and pineapples, are available throughout the year but their supply declines during certain seasons. When supply declines, market prices can more than double. During the peak seasons, the prices fall and a large amount of fruit is wasted. Processing these commodities could significantly reduce waste. There are several methods of processing fruits suitable to the small-scale producer and these vary according to the type and nature of the fruit.

Drying

Drying is the most common method for fruit processing. Bananas, pineapples, tomatoes and mangoes are the most commonly dried fruits in Uganda. Drying has several advantages. It preserves the produce. As most fresh fruits and vegetables have a high moisture content (70-95%), drying helps to reduce weight and bulk, making produce convenient to handle. Drying reduces the rate of germination in seeds and sprouting in roots, tubers and bulbs. There are two methods: open sun drying and drying in solar dryers.

Most fruits are dried using solar dryers. With this technology, fruits dry faster, products are protected from dust and insect contamination, and the overall quality is very good. There is considerable interest in solar drying for income generation. Producers have taken readily to the solar dryer technologies developed at Kawanda Agricultural Research Institute, particularly small-scale producers and those from outside Kampala district, where fruit is abundantly available. The numbers of dryers and dryer groups, as well as the production of dried fruit, have increased dramatically within the last three years. Fruit is dried in a variety of differently sized dryers and the quality of the products vary widely in terms of moisture content, shape, texture and colour. These variations are attributed to several factors, including variety of the fruit, its ripeness at the time of drying, the preparation methods and differences in the drying times due to the vagaries of the weather.

Small-scale cabinet dryers (Figure 21.1) offer the best opportunity as they require only a small capital investment, are easy to operate and manage, and can easily be replicated as groups or individuals earn money to expand production on a reasonable and manageable scale. They can handle 2 to 4kg of fresh fruit. In March 1994, a total of 66 dryers were in operation. The majority of these dryers, 85%, were of the small-scale cabinet type. Of these, 45% were based on the design developed by Kawanda Agricultural Research Institute and 55% were based on designs of other companies, mainly Fruits of the Nile. The dryers were owned by more than 50 individuals and groups, with 60% of the dryers owned by individuals and 40% by groups. Group ownership is more significant for larger dryers, 65% of which were owned by groups. Currently, over 80% of the dryers are owned and operated by women or women's groups. Other groups operating dryers include youth groups, church groups and school groups.

Figure 21.1 Diagram of a Kawanda Cabinet Dryer

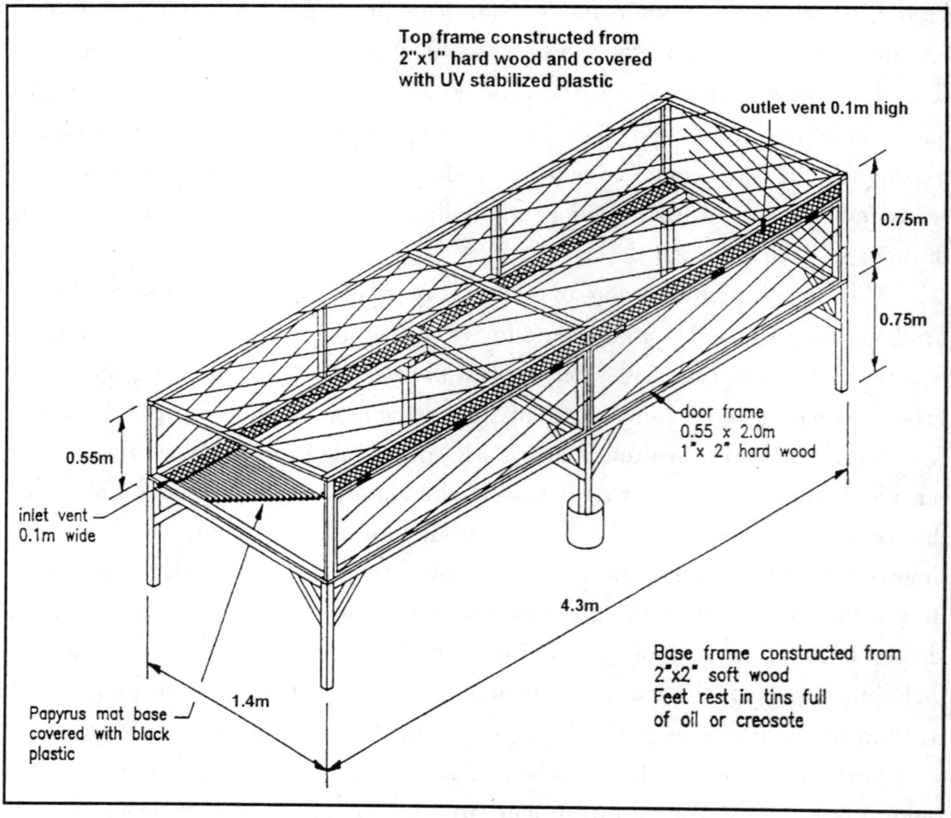

- **Fruit preparation for drying:** All equipment, such as drying trays and knives used to prepare fruit, is cleaned with a weak bleach solution. The person preparing fruit must have clean hands with nails cut, or may wear gloves. None of the fruit is given any kind of chemical/preservative treatment, thus enzymic browning is a common reaction. Fruits are washed with clean water. Pineapples are then peeled and cut into slices across the core, typically 2 - 3mm thick. Bananas are peeled and the pulp is cut into pieces, typically 5mm thick. Mangoes are peeled and cut around the stone into slices 3 - 5mm thick. Tomatoes are cut into slices 3 - 5mm thick. The pieces are laid out on trays which are then placed in the drying chamber.
- **Market for dried fruit:** At present, local market demand is very low; small quantities are sold in supermarkets where foreigners buy them. Traditionally,

Ugandans are not used to eating dried fruit. There is only one major market channel for dried fruit producers: the Ugandan-based private company Fruits of the Nile buys dried fruit, subject to certain visual quality standards, for export to the UK, Sweden and Denmark.

Juice extraction and fermentation processing

Bananas, pineapples and passion fruit are processed into juices on a small scale by individuals or groups. The juice can be consumed fresh or fermented into wines, beers and spirits. The juices are normally pasteurized and vacuum packaged in glass or plastic bottles. The bottles and caps are locally manufactured in Uganda. Corking machines, mainly wooden, are also made locally. The small-scale juice industry is limited, because the juices ferment and lose their natural flavours quickly, due to insufficient pasteurization and vacuum packaging. In addition, their stability is always a problem, particularly in the case of passion fruit juice, which separates into layers. The department of Food Science and Technology at Makerere University is researching techniques for stabilizing passion fruit juice using tamarind, which is locally available. The results thus far are highly promising because the juice stabilizes and no off-flavours have been noticed. Pectin, natural or artificial, imparts a noticeable flavour that is undesirable to consumers. The juices are mainly consumed locally.

- **Traditional juice extraction:** Bananas (beer variety) are ripened using a pit or rack with fire as the heat source. The fruit is dehanded, peeled and pulped. If the juice is for fermentation, peeling is not necessary. The juice is extracted by squeezing the pulp with spear grass in a "canoe" (a hollowed-out log). The juice is filtered, the spent pulp removed, and the juice is then pasteurized in pans at approximately 72C for 2 seconds. The hot juice is vacuum packed into bottles, which are then cooled in water and packed into crates or boxes. Passion fruit juice is extracted the same way, except that it is pulped by beating or squeezing, using polythene film. Sugar is added before the juice is separated by filtration, pasteurized in pans, vacuum packed in bottles, cooled and crated.
- **Traditional juice fermentation into wine, beer and spirits:** Banana beer and spirits earn more than 60% of the income of the rural people in the Western, Southern and Central parts of Uganda. Banana beer, locally known as *tonto*, and spirit, locally known as *waragi*, can be fermented traditionally

from banana juice. The unpeeled banana fingers are mixed with spear grass and trampled. The spent pulp is removed by filtration. The juice is fermented in "canoes" by adding roasted and milled sorghum as yeast. The result is *tonto*. The flavour of *tonto* changes with storage, as fermentation continues. Its shelf life is up to two weeks, longer under refrigerated conditions. If the *tonto* is distilled, the result is *waragi*. Bananas can also be fermented directly without prior juice extraction. They are first ripened (pit or rack), dehanded, and the fingers collected into "canoes" or drums. They are then chopped vigorously, a reasonable amount of water is added, and the mixture is covered and left to ferment for 5-7 days. Thereafter, the mixture is distilled to produce *waragi*, which is always stronger than *tonto*. This kind of processing is not as laborious as the former, since it skips the activities of banana juice extraction. The disadvantage, however, is this spirit is less desired than *tonto*, due to the ripe banana flavour characteristics that are retained.

The banana spirits, which are more than 60% alcohol, contain a lot of impurities including methanol, and they are responsible for some of the poor health conditions in rural areas, where more than 90% of the spirit is consumed. The spirit is a source of raw material for the distilling industries in Uganda; it is purified to make two beverages under the brand names "Uganda Waragill" and "Seven Hills". Traditionally, passion fruit and pineapple juices are also fermented to make wines, but this is not well documented.

Other fruit processing

- **Banana-based pancakes and doughnuts (*Mandazi*), and figs:** Pancakes and doughnuts are mainly processed from apple bananas (sweet bananas). After ripening (normally at room temperature), the bananas are dehanded, pulped and beaten or chopped thoroughly into a paste in pans. Cassava flour is added until a dough that no longer sticks on the surface of the pans is formed. Spices like hot chilli may be added. The dough is sliced into desired shapes, and the slices are deep fried. Figs are processed from the cooking variety. After peeling, 5mm slices are cut, deep fried and packaged in low-density polythene. Figs are sold mainly in supermarkets, where they are purchased by foreigners.
- **Banana flour:** Banana flour is a new technology initiated by the department of Food Science and Technology at Makerere University. The flour is made by

pounding dried banana chips with a mortar and pestle. The flour can be mixed with hot water to reconstitute the banana paste. The major disadvantage is that the original *matook* flavour is lost, and people who are used to consuming fresh bananas have no desire for the flour. It is mainly processed in the eastern parts of Uganda, where maize and millet are the staple foods.

- **Jam and marmalade processing:** Jam and marmalade are made from pineapples, guava, oranges and strawberries in scattered parts of Uganda, mainly in the Central region. Guava is mainly used, due to its high pectin content. Farmers who process these products complain that the process is labour intensive and they must compete with imported products, especially from Kenya. Their jam is not normally appealing, especially in terms of colour.

Processing Root Crops

Sweet potato and cassava are the main root crops grown and processed in Uganda. Sweet potato production has recently increased and plays a supplementary role to cassava, especially in regions where African cassava mosaic virus (ACMV) has reduced cassava production. Due to ACMV, bitter cassava varieties that are resistant have been introduced in some areas to replace the sweet, non-resistant varieties. Unfortunately, the new varieties have a high cyanide content and must be processed to reduce or completely eliminate cyanide. In addition, there has been great demand to diversify sweet potato and cassava use, in order to find alternative markets and extend their availability. Processing is the best answer to this demand. Several processing methods have been employed.

Drying

Since cassava was introduced in Uganda, open-sun air drying has been the traditional method of processing it. Sweet potato is also sun-dried, mainly in the eastern and northern parts of Uganda, but solar drying has also been introduced. Both the solar-dried sweet potato and open-air dried cassava are further processed into flour, for product development or for processing into paste. In preparation for drying, both cassava and sweet potato are uprooted, peeled, washed and cut into appropriate slices for efficient drying. The slices are put on mats, grass or roofs for open-sun drying. Fungal growth is a major problem here. After drying, the slices are stored in sacks, granaries or processed into flour.

A range of manual and powered processing machines have been developed. The machines, which include graters, chippers and pressers, are being tested in parts of Uganda where bitter cassava varieties are grown. The machines process cassava into products that can dry faster and help to increase the surface area for the enzyme linamarase to hydrolyze linamarine to reduce the cyanide content. A constraint on the use of these machines is their initial cost, which most farmers cannot afford, particularly as the price of dried cassava flour is very low and cassava is considered a poor man's food in Uganda.

Cassava fermentation

This is one of the recommended traditional methods for cyanide reduction. It is mainly practised in eastern and northern parts of Uganda, because fermentation imparts a characteristic flavour the people of these regions like. The cassava is peeled, chopped and covered with polythene for about one week, depending on the cultivar. Under these conditions, the cassava ferments. After fermentation, it is removed and sun dried. It is later stored or processed into flour. The cyanide content is reduced to safe levels.

In northern parts of Uganda, especially the Lira district, cassava is commonly fermented further to make spirits. After the fermentation process described above, the cassava is sun-dried and ground into flour. Water is added to make a paste, which is allowed to stand for five to seven days in pots or holes dug in the ground. The paste is then roasted or baked for colour development and moisture elimination. Brewing is made by mixing the product with ground germinated millet or sorghum in certain ratios, and the mixture allowed to ferment for five to seven days in pots or drums. The fermented product is then distilled into spirits called Liralira. The spirits are consumed locally, and are a source of raw material for a distilling industry in Lira district which manufactures the "Seven Hills" beverage.

Processing cassava and sweet potato

Sweet potato and cassava product development activities have just begun and are focused in the National Postharvest Research Programme at Kawanda Agricultural Research Institute. The work is done in collaboration with the National Research Institute (NRI), the International Potato Centre (CIP), the department of Food Science & Technology and the Department of Chemistry at Makerere University.

- **Sweet potato product development:** Market and consumer studies undertaken in selected areas of Lira municipality in northern Uganda, where sweet potato is largely grown, showed a potential demand for new products. Sweet potato processing technology was introduced to three women's groups, three individual processors and one industrial bakery. The products were doughnuts (*Mandazi*), chapati, bread, buns and pancakes. In all the products, 30% of the wheat was substituted with sweet potato flour or 50% with boiled and mashed sweet potato. It was found that consumers accepted all the products. Comparison of production costs of sweet potato products and pure wheat products showed that using sweet potato as a processing ingredient increased the net revenue per product piece by 12% to 68% , depending on the product and the form of processed sweet potato product used. In all products, the use of sweet potato flour gave the greatest net revenue increase. The costs of water and firewood used to wash and cook the sweet potatoes have been found to be very minor. Currently, there are five women's groups producing doughnuts (*mandazi*), chapati, buns, chips, cakes and biscuits on a small commercial scale. The doughnuts, chapati, pancakes, and crisps are deep fried. The buns, cakes and biscuits are baked in locally constructed ovens. The major constraints to sweet potato technology adoption are the seasonal availability of both fresh product and sweet potato flour, the poor quality of the sweet potato flour, the labour intensiveness of grating sweet potato, and poor product quality which results in poor sales.
- **Cassava product development**: Cassava value-added products, including cassava flour, dextrin, glue, biscuits, pancakes, buns, bread, doughnuts and baby foods have been processed by the departments of Food Science & Technology and Chemistry at Makerere University. Of these, the technology of three products has been successfully transferred to farmers in Iganga district. These are:
  - biscuits
  - composite bread made by mixing cassava flour with soya or with millet
  - baby foods composed of cassava flour and soya or composed of cassava flour, millet and soya.

  The transfer process was initiated by encouraging farmers in this district to properly harvest and store fresh cassava, a commodity that is widely grown there. Later, product development was introduced by training farmers on processing cassava into flour using the soaking method that reduces cyanide. Lastly, training on product development was done. Currently, several farmers are in big business with composite bread and biscuits, while the baby foods are mainly for household consumption to improve nutrition for children. The major constraint is the ACMV, which has reduced cassava acreage, generating a substantial increase in the price of the fresh roots. However, profits are still realized. Dextrin, glue, ethanol and soft drinks are cassava products made by the Chemistry Department, targeted for large-scale industry to adopt and process.

Processing Leafy Vegetables

Uganda grows a wide variety of leafy vegetables. The majority are consumed fresh in local markets and processing is limited. The most common processing method is drying. There are several individuals or groups involved in drying vegetables using solar dryers. Open-sun drying is also common. The majority of individuals who dry fruits with solar dryers also dry leafy vegetables. The dried leafy vegetables include cabbage, vergans (*nakati*), African spiderherb (*jjobyo*), cowpea leaves and cocoyam leaves. The process involves washing with clean water, sorting, trimming, slicing and blanching in boiling water for about two seconds. This helps to inactivate the browning enzymes and fixes the natural colour of the leaves. The leaves are then placed on dryer trays and dried accordingly. After drying, the leaves are broken into small pieces and packaged in clear low-density polythene. Some of these are sold locally to supermarkets, while others are exported to Europe.

Processing Spices

A number of spices, including hot chilli, ginger, curry, onions and garlic, are well-suited to the ecological and general pattern of small-scale farming in many parts of the country, especially Buganda, Busoga, Mbarara, Bushenyi, and Fortportal. Owing to the low level of production of various types of spices in the country, there are not many major or significant enterprises involved in spice processing. A few enterprises (Megatrends, Elgonia Industries and RECO) operate on a fairly large-scale based on the local supplies of raw material. Most processors are engaged in supplying the home market. Drying, curing and grinding into powder are the major processing methods for spices. Locally processed curry and ginger powder is found in all local markets in Uganda. Onions and garlic, after curing, are mainly packaged in plastic gunny bags and transported to fresh markets. The USAID-funded Agribusiness Development Centre (ADC), which promotes the production and utilization of high-value, non-traditional export commodities, has funded some farmer-based projects to produce and process chilli for the export market; those involved are now exporting this commodity.

Storing Horticultural Commodities

Fresh horticultural crops are highly perishable and present a very big storage problem, compared to grains and oil seeds. At peak seasons, substantial losses occur at both farm and market levels due to lack of proper storage facilities. Hence, most fresh produce ends up being sold at give-away prices. In addition, the produce is typically exposed to direct sunlight and drying winds in Ugandan market places, which accelerates shriveling, wilting and direct injury from heat and solar radiation.

There is very little storage of freshly harvested commodities at farm level. At market level, fruits not left in the open are normally stored in plastic gunny bags. Bulk storage of commodities in stalls is common. Heat and moisture build up as a result of respiration, causing fermentation and decay problems. Under these conditions, ethylene-producing commodities like tomato and passion fruit release this aging hormone, which causes senescence in neighbouring commodities. Latent infection is also encouraged. Most commodities are displayed on carts, others in rough boxes. Several commodities, including bananas, cassava and sweet potato, are put directly on the bare ground where they become soiled; they are not removed until they are sold or they deteriorate and are discarded. Their quality is highly reduced by soiling.

Some markets, especially those around Kampala, have done a commendable job of constructing permanent sheds where commodities are graded and stored. But most of these structures are wide open, so temperature, humidity and dust contamination are still a problem. The general sanitation of storage places is poor and rodents are common. Pathological problems are common with few control measures. There is, therefore, a great need to put more thought into improving storage facilities and practices.

Conclusions and Recommendations

Uganda's current postharvest systems for horticultural commodities are not well developed, but there is high potential for the horticultural processing industry. The level of its performance and contribution to postharvest loss control can be greatly improved. A number of steps would help to the improve the postharvest system.

The Uganda National Bureau of Standards should enforce the quality standards of the locally processed products, to ensure that safe products reach the market. People need to be informed about the range of processed commodities and the available processing methods. Rural Ugandans also need to be sensitized on the issue of tastes and preference; acceptability of some of these products is sometimes psychological, which may be due to lack of proper knowledge. There is a need to train extension staff specifically on postharvest systems, so that they can disseminate postharvest information to farmers. Postharvest facilities at all levels should be improved, especially the infrastructure. Through government and donor assistance, lines of credits should be established for easy accessibility of credit at low interest rates to entrepreneurs. Donor assistance to non-governmental organizations involved in fruit and vegetable processing is highly encouraged. Supply agreements between processors and growers, in which both sides are equitably protected, should be encouraged. This will help to reduce the stiff competition for particular supplies.

Stored produce should be inspected and storage facilities cleaned on a regular basis to minimize the build-up of pests and discourage the spread of diseases. When inspecting stored produce, any infected or spoiled produce should be removed and destroyed. For maximum effectiveness, storage structures should be constructed with optimum ventilation and should be insulated. Ventilation can be improved by locating air holes at the top and bottom of the structure. Open and rough storage structures should be avoided. Storage of fruits and vegetables in plastic gunny bags and in mixed loads should be avoided. Commodities should be protected from soil contamination.

The technologies of storing fresh cassava and sweet potato roots developed and tested by Kawanda Agricultural Research Institute that are suitable at the farm level should be transferred to farmers. Cassava was stored for five months in polythene bags containing moist sawdust and wood shavings, with no physiological or microbial deterioration, although the roots became sweeter. Sweet potato roots were stored for eight weeks in ventilated and non-ventilated clamps and pits, and showed minimum weight and quality losses. Many of the more perishable fruits and vegetables, however, simply cannot be stored for any appreciable length of time without refrigeration. These commodities must be marketed as soon as possible, or be processed into products with a longer shelf life.

References

Devereau, A.D. and Bockett, G.N.N. 1996: *Sweet potato Storage - Is there a need to improve traditional practice in Uganda.* Kampala: Kawanda Agricultural Research Institute, Postharvest Programme Publication.

FAO. 1985: *Prevention of Postharvest Food Losses:* A training manual. Rome: UNFAO.

Greely, M. 1993: *Postharvest Technologies: Implications for Food Policy Analysis.* EDI Development Policy Case Series, Analytical Case Studies No. 7.

Kaaya, N.A. 1995: Postharvest Technology in Uganda: Problems and Possible Strategies for Horticultural Crops. In: *Sustainable Agriculture Research In Uganda.* Kampala: ARTP/NARO.

Kader, A.A. (ed.). 1992: *Postharvest Technology of Horticultural Crops*, 2nd Ed. California: University of California, Division of Agriculture and Natural Resources. Publication 3311.

Kapasi-Kakaza, J. 1996: *The Food Processing Industry in Uganda: Importance, Constraints and Prospects.* Paper presented at the Inter-University Council for E. Africa Food Science Workshop, Arusha, Tanzania, 9-13 September.

Kyokunda, G. 1994: *The New Uganda Trade and Investment Opportunities, 1st Ed. Sector Profile: Food Industries.* Kampala: UIA.

Owori, C. 1996: *Solar Drying as a Business.* Paper presented in a Training Workshop on Postharvest Technology of High Value Crops for Agricultural Extension Personnel, Kampala, Uganda, 16 January.

Owori, C. 1997. *Sweet Potato Product Development in Lira Municipality, Uganda.* Kampala: Kawanda Agricultural Research Institute, Postharvest Programme Publication.

Slim, M.N. and Odong, N. 1996: *Fresh cassava roots storage in different level of wet storage media.* Kampala: Kawanda Agricultural Research Institute, Postharvest Programme Publication.

CHAPTER 22

ENVIRONMENTALLY FRIENDLY POSTHARVEST SYSTEMS: PINEAPPLES, CITRUS FRUITS AND ROOT CROPS IN MALAWI

by H.F. Mbeza and I.J. Minde

INTRODUCTION

Processing changes a commodity's form so that handling, transportation, consumption or storage are made easier. In some cases, processing reduces the non-economic weight so that the product value per unit of weight is increased. Processing for consumption refers to transformation of the product, including the separation of edible and non-edible parts. Processing for storage may involve physical and/or chemical transformation of the product in order to increase the shelf life. Processing sustains a marketing or consumption system when demand is continuous and production is seasonal. Moving a product in and out of storage ensures that at least some supplies are available throughout the consumption period. In the smallholder sector, particularly households with near-zero food surplus, the desire to process and store in order to sell may be less important than the concern to ensure adequate food supplies for the family in the off-agriculture season. This paper presents data on three processing challenges facing small farmers in Malawi: the processing of fruits, the problems of drying of cassava chips and treatment of sweet potato for pit storage.

PINEAPPLES AND CITRUS FRUITS

Malawi has a wide range of climatic conditions and soil types which allow a variety of fruits to be produced (MOALD, 1994). Fruits are a valuable source of vitamins and mineral salts which are essential for body protection against disease. Some fruits also provide proteins, carbohydrates and oils. The fruit crops grown are bananas and plantain *(Musa spp.)*, mangoes *(Mangifera Indica)*, pineapples *(Ananas comosus)*, avocado pears *(Persica americana)*, papayas *(Carica papaya)*, apples *(Malus domestica Bork),* peaches *(Prunus persica),* plums *(Prunus domestica),* pears *(pyrus coousis),* tangerines *(Citrus reticulata),* grapefruits *(Citrus paradisi),* lemons *(Citrus aurantifolia)* and oranges *(Citrus sinensis)* (MOALD, 1994). These are chiefly seasonal, with mangoes peaking in December-January, tangerines in May-June, pineapples in July-August.

There are two main uses for these fruits: direct consumption and processing.

When directly consumed, individuals suck out the juice, discarding the peel and the flesh, contributing to sanitation problems in urban areas. The peel and flesh of pineapples and citrus fruits make up a significant proportion of the fruits' weight and nutrients. These heavy fruits are normally transported long distances to markets. If fewer whole fruits and more juice products were transported to urban areas, both transport costs and urban sanitation problems would be reduced (Mbeza, 1995).

Malawi's one fruit factory, Mulanje Peak, is in the Mulanje district of the Southern Region. The factory was built there because it is the main growing area for pineapple, the main fruit processed. Other fruits, such tangerines, are not processed at the factory because they are grown 170 kilometres away in Mwanza district. For juice production, it would be advantageous to have a primary processing mechanism at the farm level or near the production area before transporting the product for further processing at the factory. This would be beneficial because transporting the "crude" juice is cheaper than transporting the heavier whole fruit. Because transport costs are factored into the selling price, lower transport costs would lower the price of the final juice product.

When fruit is processed in a factory, the peel is removed and the fruit is sold as rings or squeezed into juice which is either canned or bottled. The disposal of the peel, seeds and water waste in processing presents problems; they must be transported from the factory, increasing the cost of the juice or rings.

Furthermore, if the fruit were squeezed on the farm, the residues or waste would remain and could be further processed into compost manure or animal feed. This would enhance environmentally friendly agriculture, as less or no chemical fertilizers would be used or fed to animals.

Figure 22.1 The *Kufinya* Juice Press

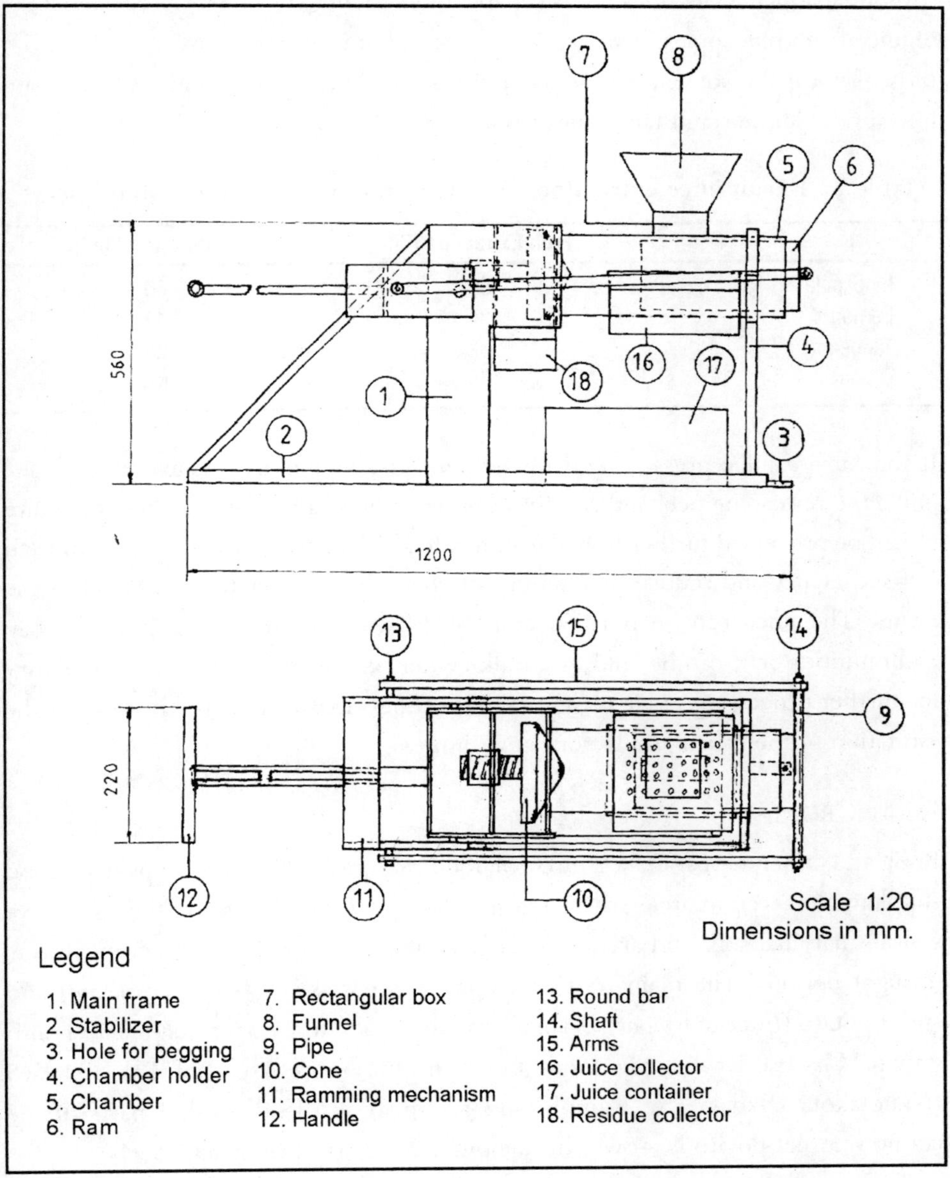

The Kufinya juice press

The *Kufinya* Juice press (Figure 22.1) was designed at the Bunda College of Agriculture Engineering Department (Mbeza, 1996). The total weight of the machine is 40.5kg. It can be transported easily on a bicycle and dismantled easily without spanners. The machine is capable of extracting 50%, 34%, 32% and 35% of juice from pineapples, lemons, tangerines and oranges respectively (Table 22.1). It is also capable of extracting the juice at 60 l/hr, 42 l/hr and 28 l/hr from pineapples, lemons and tangerines respectively (Mbeza, 1996).

Table 22.1 Fruit juice extraction efficiency and rate using the Kufinya press

	Fruit Extraction (%)	Rate (litres/ hr.)
Pineapple	50	60
Lemon	34	42
Tangerine	32	28
Orange	35	NA

If the *Kufinya* juice press is used in the fruit production area to extract "crude" juice, the remaining peel and residues can be incorporated into the soil as manure or can be processed further to make animal feed. This can enhance long-term food self-sufficiency and reduce the amount of chemical fertilizer used on most of the farms. The juice can be purified and sterilised before transportation to urban communities or it can be sold to a bulk center before transportation to a factory for further processing, such as canning or bottling. This would greatly improve the sanitation of the urban and factory environment.

Root Crops

Root and tuber crops are a source of food and cash. They are important food supplements, even in areas where maize is a staple food. Roots and tubers survive well in marginal soils and are particularly important as food security crops during drought periods. The main root crops grown are cassava (*Manihot esculenta*) and sweet potato (*Ipomoea batatas*). A small amount of potato (*Solanum tuberosum*) is also grown. Cassava is a staple food grown in the shore areas of Lake Malawi (Nkhotakota, Nkhata Bay, Rumphi and Karonga) and less in other parts of the country. Sweet potato is grown throughout the country (MOALD, 1994).

Generally, cassava and sweet potato are environmentally friendly crops. Both prevent wind erosion. During the windy dry season, two-meter tall cassava plants protect the soil against wind blast. Sweet potato has a netting effect on ridges that prevent soil erosion (Benesi, Minde, Nyondo & Trail, 1995). Pesticide and herbicide contamination of the soil and water is not a consideration, as none are used by farmers. Both cassava and sweet potato lend themselves well to diversification. While both are planted as monocrops, much intercropping is practised by farmers using various cropping systems. Some new trials involving cassava and sweet potato combined with agroforestry practices are under way in Malawi. As their nitrogen requirements are less than those of other crops, farmers do not use commercial fertilizer on cassava or sweet potato. Farm surveys in 1995 (Benesi *et al.*, 1995) showed that farmers in northern Malawi use limited amounts of cassava peelings mixed with animal manure to fertilize their cassava.

Cassava processing

Cassava can be stored in dried form. One method is to peel and chop cassava before drying. The traditional way of chopping cassava in Malawi produces large chips which are not uniform in size. Solar energy is used for drying and this does not ensure a uniform drying rate. The large chips normally dry outside, resulting in mould development or discolouration inside.

The International Institute of Tropical Agriculture (IITA) developed a cassava chipper which was tested in Malawi (Kazembe-Phiri, 1994). The working rate of the chipper was higher (90.2 kg/hr) than that of hand chipping (20 kg/hr). However, hand chipping resulted in lower loss (5%), compared to the machine (10%). Despite the loss, the smaller machine-made chips result in a faster, more uniform drying rate, with all the chips drying at or near the same time. It is likely that the dried machine-made cassava chips would develop less mould in storage and would be of higher quality for later consumption.

Sweet potato storage

Sweet potato is the world's sixth most important crop. In Africa, about 1,714,000 hectares are under sweet potato cultivation and the production is about 5,474,000 tons per year, or approximately 4% of the world's production (Burton, 1996; Lenne, 1991). Countries in Eastern and Southern Africa produce 76% of Africa's total sweet potato production (Roots, 1994).

A major constraint to sweet potato production is the smallholder farmer's lack of suitable and cheap storage facilities. In the tropical world, the sweet potato is most commonly stored in pits. Pit storage has been reported in Indonesia, Zimbabwe and Malawi. The pits are dug on open ground and ash is often sprinkled on the bottom and sides of the pit and between layers of the tubers. The pit may be covered by grass and soil, a stone and soil, or grass with a shed or woven mat (Woolfe, 1992, Kwapata, 1984). Golob and Webley (1980) report widespread use of ash for insect control in stored products in developing countries. Traditionally, ash is applied to control insects and some fungal diseases responsible for root decay (Kwapata, 1984). However, Kwapata (1984) also reported that the application of ash in the traditional pit method does not significantly extend the storage life of the sweet potato (Njazi-Tembo, 1994).

Trials conducted by Mbeza, Minde and Mjiku (1997) on sweet potato storage in pits suggest that alternative treatments do affect the length of storage. Tables 22.2 and 22.3 indicate *Kakoma* (TIS 3017) and *Kenya* (SPNO) varieties of sweet potato stored in pits at Bunda College. The treatments included a chemical (Actelic), *dambo* sand, maize and bean stover ash. The control was storage without treatment. For both types of sweet potato, sand treatment was the best, as it caused the least weight reduction over time. Maize stover ash and bean stover ash resulted in the largest weight losses. This experiment suggests that the use of ash should be discouraged, while *dambo* sand should be encouraged. Although the Actelic treatment showed encouraging results, the effect of these chemicals on the environment may not be positive.

Table 22.2 Weight retention of stored sweet potato (Kakoma)

Month	Actelic %	Sand %	Maize Ash %	Bean Ash %	Control %
June	100	100	100	100	100
August	53	64	59	52	54
September	45	52	50	43	46
October	38	46	50	30	38
November	7	42	39	23	34

Table 22.3 Weight retention of stored sweet potato (Kenya)

Month	Actelic %	Sand %	Maize Ash %	Bean Ash %	Control %
August	100	100	100	100	100
October	50	58	40	38	41
November	41	52	30	28	29

Research Needs

Multi-disciplinary focused research by a team of both biophysical and social scientists is needed to conduct continuing farm-level participatory research on post-harvest systems. Further information is required on disposal mechanisms for citrus and root crop waste, the quantities of such waste disposed by households, the chemical composition and nutrition value of the waste, and the feasibility of generating economically sound, socially and culturally acceptable and environmentally friendly products from these crop wastes. These products will go a long way towards increasing total farm productivity, family cash incomes and promoting environmental sustainability.

References

Benesi, I. R. M., Minde, I.J., Nyondo, F. and Trail, T. 1995: *Adaption Rate and Impact Assessment of Accelerated and Distribution of Cassava and Sweet Potato Planting Material as a Drought Recovery Measure in Malawi.* A report prepared for SADC/IITA/ SARRNET, Lilongwe.

Burton, W.G. 1996: *The Potato: A survey of its history and factors influencing its yield, nutritive value, quality and storage.* 2nd Edition. Wageningen, Holland: H. Veenman and Zonen N.V.

Golob, P. and Webley, D.J. 1980: *The use of plant materials as traditional protectants of storage products.* U.K.: Tropical Products Institute. Report GL30.

Kazembe-Phirl, H. W. C, 1994: *Performance and evaluation report of IITA peeling knives and chippers for cassava processing.* Chitedze Research Station, Malawi.

Kwapata, M.B. 1984: Effects of sun curing time and application of ash on the storability of sweet potatoes. *Research Bulletin Bunda College of Agriculture* 1983-84 12:171-180.

Lenne, J.M. 1991: Diseases and pests of sweet potato. *Natural Resources Institute Bulletin* No. 46. Natural Resources Institute.

Mbeza, H. F. 1995: Small Scale Juice Extraction in Malawi: A dream or reality. Paper presented to the seminar on "Improvement on Agriculture Production and Food Security in Malawi". 14 to 16 December, 1995. Kalikuti Hotel, Malawi.

Mbeza, H. F. 1996. Design and construction of a hand operated juice extractor for mangoes, pineapples and citrus fruits by small/medium entrepreneurs in Malawi. Paper presented to the annual horticultural project meeting of the

Department of Agricultural Research (MOALD). 9-13 September 1996, Mlangeni Holiday Resort, Mangochi, Malawi.

Mbeza, H.F., Minde, I.J. and Mjiku, E.K. 1997: Pit storage of sweet potato and its economics in Malawi. Paper presented at the 4th Triennial African Potato Association 23-28 February, 1997. Pretoria, South Africa.

Mbeza, H.F., Minde, I.J., Kwapata, M.B., Mjiku, E.K. and Nyirenda, K.S. 1997: Storage of sweet potato in Malawi. 2nd year report (unpublished).

MOALD (Ministry of Agriculture and Livestock Development) 1994: *Guide to agricultural production* Lilongwe: Agricultural Communication Branch, MOALD.

Njazi-Tembo, A. 1994: *Effects of ash in the traditional pit storage of sweet potato.* B.Sc thesis, Bunda College, Malawi

Roots, 1994: Newsletter of the Southern Africa Root Crops Research Network (SARRNET) and the East Africa Crops Research Network (EARRNET).

Woolfe J.A. 1992: *Sweet potato, untapped food source.* Cambridge: Cambridge University Press.

CHAPTER 23

BOTTLENECKS TO THE DISSEMINATION OF ECOLOGICAL AGRICULTURE IN KENYA

by Yucabeth M. Ogega

INTRODUCTION

Like many developing countries, Kenya has not been spared the negative environmental and economic consequences of conventional agriculture associated with the "green revolution". A great majority of the smallholder farmers in the local setting rely upon the existing extension services to support their farming activities. Such services are provided by the government and, to a lesser extent, by non-governmental organizations and the private sector.

THE BOTTLENECKS

In Kenya, the most notable bottlenecks to the dissemination of ecological agriculture can be viewed under the following categories:

Lack of local expertise on ecological agriculture

A lot of work and documentation has been done on conventional agriculture. This is not the case with ecological agriculture. Moreover, most of the agricultural technicians and professionals within the extension service have been trained and have experience in conventional agriculture.

Existing government policies

Current government agricultural policies tend to favour conventional agriculture as compared to ecological agriculture. It is no wonder then that the available agricultural training and research institutions have tailored their training programmes to conventional agriculture. Little emphasis has been placed on research dealing with ecological agriculture.

Government incentives for cash crop production

The government has continuously encouraged and provided incentives for the increased production of cash crops. This has called for the increase of land under cultivation, which has not spared the river banks, hilltops and other natural reserves that act as catchment areas. Very little attention has been given to the adoption of sound ecological agricultural practices.

Application methods and technologies

Certain methods and technologies of ecological agriculture are quite labour-intensive, particularly in the case of field crops. For example, most farmers would find it very cumbersome to apply farmyard manure to large areas of cultivated crops. In such circumstances, they prefer to use conventional methods. The same applies to the bio-intensive gardening methods of organic farming.

Lack of support and recognition of farmers' indigenous knowledge

Farmers are usually provided with technology recommendations, some of which are not compatible with local conditions and cultural values. Moreover, the farmers' knowledge is not sought before such technologies are packaged. Many farmers have come up with very good ecological agriculture innovations, but as these are not recognized and supported, dissemination of these ideas is greatly hampered.

Rapid population growth and land fragmentation

Due to increased population growth, there is great pressure on the land and this leads to continuous cultivation. Thus, certain aspects of ecological agriculture, which were previously practised and are environmentally sound – such as fallow farming – are not considered practical today. Increased population leads to land fragmentation. Under the traditional land inheritance system, the father subdivides his land to all his sons. This has resulted in small strips of land which run downhill; these are constantly under cultivation and often ploughed, which encourages excessive soil erosion.

Proposals for Solutions

Extension and training

There should be concerted efforts to introduce farm education and awareness programmes that centre on the benefits and advantages of ecological agriculture over those of conventional methods. For example, manual application of farmyard manure can be very cumbersome, but the long-term effects are worth the labour expended. Farmers can be sensitized to a variety of new techniques through local dramas and songs, as these are very common and effective in conveying messages. The technical staff and all personnel involved in agricultural production need to be trained in various aspects of ecological agriculture. This will equip them with much needed technical know-how and expertise for the success of ecological agriculture. Textbooks and relevant extension education materials must be made readily available so that information can easily be sought by anybody wishing to practice ecological agriculture. Extension materials would include posters, slides and videos.

There should also be forums where extension officers can exchange ideas and experiences. For example, when we had a Train and Visit Programme, there were very good results because the extension staff met fortnightly and shared the information gathered during visits and interaction with the farmers.

Policy reform

A shift in existing government policies towards the promotion of ecological agriculture is necessary. Government agricultural institutes should accommodate and give due recognition to ecological agricultural systems in their training curricula.

In addition, appropriate restrictions on cash crop production could be introduced and followed. For example, in Kenya's sugarcane growing areas, farmers initially did not plant cane on more than a half of their land. However, as time went by, farmers put most of the land under sugarcane because it earned good money. They were not strictly monitored.

In contrast, agricultural lending institutions are very strict with their policies. For example, before granting a loan, their agricultural staff visit the farm to-ensure that what the farmer has said on the application form is true and the whole family is in favour of the farm plan. In this way, they have been very successful.

Encourage indigenous knowledge

Farmers' indigenous knowledge should be supported, encouraged and, if possible, rewarded. When farmers are given a chance to experiment on their own, in most cases, they come up with better results that are compatible with the local conditions.

Control land fragmentation

Population control measures must continue to curb rapid population growth. Of late, there has been a small reduction in the population growth rate from 4.1% to 3.8%. In order to avoid land fragmentation, it is necessary to discourage the creation of excessively small plots. In Kisii, we now insist that land should not be subdivided to portions that are not agriculturally economical. We advise that if land is not adequate, the solution is not to create strips running from the hill tops to the river – which is the common feature in the Kisii Highlands. The land can be subdivided along the contour lines, so long as there is assured access to water.

Conclusion

In conclusion, ecological agriculture has been and is being practised in Kenya, but on a very small scale. Much more would be achieved if concerted efforts were put into the recognition and promotion of ecological agriculture by the governmental and non-governmental institutions dedicated to promoting the agricultural sector.

Chapter 24

The Kasisi Agricultural Training Centre, Lusaka, Zambia

by Paul Desmarais, S.J.

Background

The Kasisi Agricultural Training Centre (KATC) is a church-sponsored project located near the Lusaka International Airport. KATC has been functioning for 23 years as a farmer training centre for young families. The staff consists of seven agronomists, one biologist, one engineer, one forester and four blacksmiths. One staff member has done a course on co-operatives in Antigonish, Canada. All have received some specialized training in organic farming. In 1982, an Appropriate Technology Workshop was established. An agroforestry programme was started in 1986 and expanded in 1988. Six years ago, a training programme in ecologically sustainable agriculture was started for grades 9 and 12 school leavers. In January 1997, short courses in organic farming were introduced.

Kasisi Agricultural Training Centre Programme Activities

Family training

At the centre, residential training is provided for up to ten families. During their two-year programme, the trainees learn about rain-fed crops, irrigated vegetable production, marketing and cattle management. The training is very practical and informal – the fields are the classroom. Each family has its own area for growing vegetables and rain-fed crops. There are no scholastic entry requirements. The trainees receive an initial loan for inputs, which they repay with the income from produce sales. They also use the income for their upkeep and are usually able to purchase oxen, ox implements and hand tools to be taken home at the end of the training period. Results are measured by the production on their plots, by the co-operation amongst themselves and within their families and by their purchase of farm equipment and cattle.

Informal school

Six years ago, we began a training programme in ecologically sustainable agriculture for grades 9 and 12 school leavers. This is a two-year, non-residential programme for 10 to 20 boys and girls. The training method is again production oriented and practical, but also includes theory because these young people have had some secondary schooling. The young people are taught how to produce more food on small farms, while improving soil fertility and conserving a healthy environment through the use of organic fertilizers such as compost and green manure, careful crop rotation, and recycling of materials. They gain practical experience in regenerative agriculture (organic farming and gardening), agroforestry (improved fallow, living fences, windbreaks, indigenous and exotic trees for fruit, vegetables, insecticides, and firewood), conservation tillage, and integrated pest management.

Short course training

Short, three- to five-day training courses for small-scale farmers, agricultural extension officers, community-based leaders and primary school teachers were started in January 1997. A set of sustainable agriculture training modules have been developed in the nine courses that have been given to date.

Appropriate technology

In 1982, KATC established the Workshop for Appropriate Technology to do research and development of equipment and tools suitable for production and use in rural areas. These include ox carts, treadle water pumps, an ox-powered grinding mill, improved neck yokes and harnesses for oxen, a hand-operated oil press, sisal cement roof tiles, wheelbarrows, small weigh scales and a biogas digester. Research and development has been undertaken on passive solar power and solar box ovens. A solar box oven was built. A women's group in the village is using one model for baking scones and they like it. Another group tried the solar box oven, but didn't like it because using a solar box oven for baking scones took longer than using firewood.

Treadle water pumps have been built for the past ten years at KATC. Initially farmers were not buying these pumps. Some brave farmers bought a few and found they work well and are economical to run. Once a few people began using them and found that they could irrigate the vegetable gardens cheaply, others followed. The aim of most farmers had been to save their money to buy a diesel or petrol water pump in order to plant more vegetables. Three years ago, even the students at the centre used a diesel water pump. They didn't make any profit from the sale of their vegetables – all the income went to buy oil and diesel. During the last two years, the students have used treadle water pumps and they were able to realize a profit from the vegetable gardens. One farmer recently approached us to buy a treadle pump. He has a petrol pump, but sees his friends irrigating with treadle pumps – without the fuel costs. He sees he will make more money with a treadle pump than with a petrol pump, without the hassle of getting fuel and lubricants from town.

A steam-powered water pump is being built, again using the sun via a parabolic reflector to heat the water to steam. The steam will drive the pump.

The workshop also trains young people in iron-smithing and basic carpentry. In 1995, we offered a three-month blacksmith course to nine local school leavers. The idea was they would set up small workshops in their villages, assist in the repairs of local farmers' ox implements and earn extra money. Unfortunately, not one has gone back to the village. Young people want training in order to find a permanent job in Lusaka.

A biogas plant is providing lighting and cooking at two houses. The biogas plant is based on an Indian design with a floating dome. Ox power has been used to run a maize grinding mill and a water pump (ox-powered water pump Bidon Verseur).

Agroforestry

Interest in agroforestry developed in 1985, and the following year we started by planting trees for alley cropping. In 1986, we expanded by planting four other species of leguminous trees and shrubs. The trees we tried were *Leucaena leucocephala, Senna siamea, Sesbania sesban* and *Cajanus cajan*. We have, however, abandoned this approach because the rainy seasons are too unpredictable in our area. Most years, the rains are not sufficient to grow a maize crop and a tree crop. The Leucaena trees were difficult to establish in this particular field because they were severely attacked by termites, but the other trees and shrubs were easily established.

We are also using trees for live fencing and experimenting with other trees for other uses. The Department of Forestry has seconded a staff member who works at KATC and in the villages. In the last four years, small-scale farmers have been seeking advice on trees. Until then they did not see the need to plant trees, but they are now without firewood. As well as asking what trees will grow quickly for firewood, some are planting for future roofing material. Planting of fruit trees is still not much practiced.

Other activities

- **Animal husbandry:** The project keeps cattle and pigs, for income generation and for training purposes. In the last few years, cattle have also been raised to provide draft animals for sale to trainees and villagers. In 1986, the project started breeding improved dairy cattle and began milking them in 1992.
- **Agricultural extension:** KATC is also involved in extension work in the surrounding villages. The extension officer visits graduate families, other interested farmers, youths and, at the moment, three women's clubs. To a limited degree, seeds are also provided. Through this extension work, KATC is in contact with 1,000 families in the Kasisi area.
- **Local verification trials:** Results from government research stations, especially those of new varieties of cassava, soyabeans, improved pasture grasses and legumes, have been tested at the Centre for local adaptation.
- **Resettlement farm:** Since 1995, twelve families who graduated from KATC training have settled on an independent farm located 10 km from KATC. The graduates wish to set up a co-operative in their community.
- **Newsletter:** For the past year and a half, we have produced a monthly newsletter about sustainable agriculture for the local community.

Transition to an Ecoagricultural Philosophy

Ecoagriculture is based on broader philosophical premises than those of conventional agriculture. As well as economics, the social and cultural aspects of people's lives are valued. Ecoagriculture flows from a new vision of the world. The transition is difficult for some people to achieve.

The shift to growing vegetables organically at KATC took place in 1990. For the first year or two, we sometimes saw trainees still use synthetic fertilizer and pesticides. Gradually they saw that they could grow vegetables without adding fertilizer. There was resistance at first to making compost. There were many excuses: too little time, manure was too far away to be collected with oxen, insufficient water. However, over time the trainees have been able to market their vegetables more easily in Lusaka because they are organically grown. Marketeers say they prefer the trainees' vegetables because they have a longer shelf life. This marketing success has been a plus in getting the trainees to accept organic methods. We have not marketed our produce as organic, although some people think that we could sell it in Lusaka at specialty shops.

The instructors, who have gardens very near to the trainee gardens, were very slow in accepting organic methods and some may still be using some purchased fertilizer and pesticides. In the last two years, however, they also have started making compost. One of our instructors said he very strongly resisted organic farming methods at first. He had been brought up on a farm where it was considered valuable to produce a big crop of maize using fertilizer. The family felt they were moving away from subsistence farming to commercial farming and they valued only the economics of cropping. Now this young man sees that it is important to value the soil and the trees in order to offer his children an improved environment. He understands that his thinking has changed. That is a very big step in moving towards a sustainable agriculture.

A big philosophical shift must take place when moving from conventional agriculture to organic. The rain-fed crops at the Centre are still fertilized with purchased fertilizer. But in the last four years, more sunhemp has been planted in rotation and workers and trainees now see that the result is a much better maize yield even though they are using less bought fertilizer.

We have not stressed organic farming in the villages until this year. Since January, we have had three-day workshops with 120 farmers in the area around Kasisi. The farmers seem very keen to use green manure crops, rotations, and new crops. Fertilizer, which was subsidized until two years ago, is now unaffordable and is not always available. The farmers see that by using sunhemp, a good crop of maize can be grown with little or no fertilizer.

A lot of vegetables are grown in the villages around Kasisi for the Lusaka market. Farmers had bought chemicals in town to control insects and diseases. Now they use cattle dip chemicals to spray on their vegetables because they say the other chemicals don't work. They are very interested when we tell them that vegetables can be grown without spraying with bought chemicals. We tell them about various sprays they can make themselves. A very effective spray is made from Tephrosia, a local plant they can grow in their gardens. The farmers want to learn about these home-made sprays.

In January 1997, short courses in organic farming were introduced. These are residential courses of three to five days. To date, we have had nine courses with 20 participants each. They reside at KATC during the course. The participants are small-scale farmers (men and women), agricultural extension officers and primary school teachers who teach agriculture and supervise production units. The farmers in the three-day workshops are much more enthusiastic about organic farming than the agricultural extension officers. During the workshops, the farmers share different methods they themselves have tried and they have much more knowledge about effective methods than do the agricultural extension officers.

We have had six five-day workshops with agricultural extension officers. They have a lot of resistance to accepting organic farming, saying farmers will not accept the ideas because they are too labour-intensive. For example, one group had a hard time making a small compost pile and said farmers would not make compost piles. In fact, the farmers are quite ready to make compost piles. For them it is a great saving, as they do not have to purchase fertilizer. It is interesting to note the agricultural extension officers' perception of what the farmers will accept. Perhaps it is because the extension officers are younger people with limited practical knowledge and they have been trained scientifically in conventional agricultural methods. They might not be ready to let go of this knowledge and learn from a farmer. Ironically, we have had agricultural extension officers who have studied sunhemp, but have never seen the plant and so can't identify it in the field.

Factors affecting change

Experience at KATC suggests that people need to undergo a change of attitude and understanding if the adoption of an ecological approach to agriculture is to take place. Several factors influence the adoption of new agricultural methods:

- **Farmers' perceptions:** The farmers must recognize the need for a particular practice. For example, in 1974 tree seedlings had been grown at KATC, but we couldn't give them away to the villagers – nobody wanted to plant a tree. The villagers' attitude was that God has always provided them with trees and they would always have sufficient firewood. It was only when firewood became scarce that interest in tree planting grew.
- **Political and economic context:** The political and economic situation can help create an atmosphere that is conducive to innovation and adoption. With

the liberalized economy, it is now very difficult for small-scale farmers to market maize and they are forced to look at other crops. This helps encourage crop diversification. With the removal of subsidies, fertilizers have become too expensive and farmers are looking for alternatives. The economy has made many families poor. Families are now interested in planting traditional crops such as cowpeas for home consumption and this strengthens household food security.

- **Land tenure:** The land tenure system will also have an effect on the farming system. Traditionally, farmers could shift to another piece of land when the soil became exhausted. Now, with a growing population, there is less free land, so it is more difficult to keep shifting. Also, government is encouraging people to have title deeds to the land, which reduces the willingness to move and increases the desire to sustain the productivity of the land.
- **Information:** Farmers need information about organic production. Once they hear about the concept, they are quite willing to try it.
- **Appropriate technology:** A set of technologies has to develop in parallel with biological research. Low external input agriculture usually means more labour. Appropriate tools have to be developed for small-scale farmers to ease the workload, ensure that the work is done in a timely fashion. More efficient use has to be made of ox power as a source of energy.

Conclusion

There is now a growing interest in organic farming, especially among small-scale farmers. Even though the interest stems mainly from economic motives, farmers are now actively seeking new technologies and new crops. It is an opportune time for ecoagriculture to take hold. The Kasisi Agricultural Training Centre has made the transition and will continue to inform local farmers about ecoagricultural alternatives through its training and extension activities in Zambia.

[illegible] regional economic status [illegible] problems [illegible] [illegible] and [illegible] to look at others [illegible] [illegible] the removal of sand [illegible] [illegible] for stones and termites, [illegible] looking for [illegible] [illegible] [illegible] more than the [illegible] [illegible] [illegible] [illegible] [illegible] [illegible] [illegible] or [illegible] consumption [illegible] [illegible] food [illegible].

- Long term [illegible] the land tenure system will [illegible] how [illegible] [illegible] the land [illegible] [illegible] [illegible] could [illegible] [illegible] [illegible] [illegible] [illegible] [illegible] [illegible] [illegible] [illegible] [illegible] [illegible] [illegible] [illegible] [illegible] [illegible] people to have [illegible] [illegible] [illegible] [illegible] [illegible] [illegible] and increase the [illegible] [illegible] the [illegible] [illegible].
- Information: [illegible] [illegible] [illegible] about [illegible] [illegible] that [illegible] [illegible] [illegible] [illegible] [illegible].
- Appropriate technology: [illegible] of technology [illegible] [illegible] and [illegible] [illegible] [illegible] [illegible] [illegible] [illegible] [illegible] labour [illegible] [illegible] [illegible] [illegible] the [illegible] [illegible] [illegible] [illegible] [illegible] [illegible] [illegible] to [illegible] [illegible] [illegible].

CONCLUSION

There is [illegible] [illegible] [illegible] [illegible] [illegible] [illegible] among small-scale farmers, [illegible] [illegible] [illegible] [illegible] [illegible] [illegible] [illegible] are [illegible] [illegible] [illegible] [illegible] [illegible] [illegible] [illegible] [illegible] [illegible] [illegible] [illegible] [illegible] [illegible] [illegible] the [illegible] and [illegible] [illegible] [illegible] [illegible] [illegible] [illegible] [illegible] [illegible] and extension [illegible] [illegible].

Chapter 25

Lipangwe Organic Manure Demonstration Farm, Malawi

by Jailos J. Kanjanga

Introduction

Malawi is a land-locked country. It covers an area of about 119,140 square kilometres, of which 20% is water, dominated by Lake Malawi which runs along the eastern side of the country. Malawi has one growing or rainy season, with rainfall varying from 800mm to 2500mm. The current population is estimated at 11 million, with a growth rate of 3.3% per annum. Malawi's Gross Domestic Product (GDP) per person is estimated at US$250. Economic activity is dominated by agriculture, which employs 85% of the active population and contributes 37% of the GDP. Agriculture also contributes 90% of the Gross Domestic Export earnings.

The agriculture sector consists of estates and smallholders. The estates produce cash crops with a high endowment of inputs. The main cash crops are tobacco, tea and sugar. Substantial agricultural growth has been realized in the estate sector (5.6% per annum). The smallholders mainly produce food crops and their rate of growth has been much lower, at 0.7% per annum.

The smallholder sector operates under customary land tenure systems and is primarily subsistence oriented. As such, a large portion of smallholder land is put under food crops with maize, the staple food, on 76% of the total cropped area. Other food crops include cassava, groundnuts, rice, sorghum and millet. Arable cultivable land is dwindling every year because of the high population growth rate. Statistics show that over 56% of smallholder farmers cultivate less than one hectare of land, 31% between one and two hectares and only 14% cultivate more than two hectares. This intensive cultivation on small pieces of land results in land degradation. The smallholder sector also faces the problem of high prices for agricultural inputs and little or no access to credit facilities. Studies have shown that most of the credit issued by the Ministry of Agriculture reaches only the already progressive farmers who own collateral. To attain good yields, these farmers apply chemical fertilizers to their fields. For resource-poor farmers, fertilizer prices are escalating beyond their reach. Sustainable agricultural practices are an effective substitute for the use of such expensive conventional inputs.

Lipangwe Organic Manure Demonstration Farm

Lipangwe Organic Manure Demonstration Farm (LOMADEF) began in 1993 in Ntonda, Ntcheu district, in the southern part of the central region of Malawi. The farm is run by 17 farmers, ten women and seven men who are resource poor. For a long time, these farmers have been trapped in the bondage of low production due to escalating input prices and deteriorating soil fertility. The demonstration farm has five hectares under cultivation. Our mission is to popularise the benefits of chemical-free foods on health and the environment. The idea of establishing an organic demonstration farm was initiated by some of the current farm members, who had previously attended forums and workshops on sustainable agriculture techniques and agricultural problem analysis. These were organized by the Christian Service Committee, a local non-governmental organization.

Project Activities And Outputs

We began with maize production and other crops were added over time. The crops grown on the farm now include maize, beans, soya beans, groundnuts, pigeon peas, cassava and sweet potatoes. The crops are grown in mixture, as well as in rotation. Legumes are grown intensively to fix nitrogen and improve soil fertility. We have been able to improve the yields in our area with the use of organic inputs, organic fertilizer and organic pest control. The result is healthy, chemical-free produce.

Kraal and compost manure are used on the farm. Members conduct on-farm experiments to assess the effectiveness of manure application as opposed to fertilizer application. Members are also testing the effects of different manure application methods on maize yield. The application methods being tested are dollop, banding and spreading before making ridges. In both the 1993/94 and 1994/95 cropping seasons, Lipangwe farmers observed that their maize performed much better than the maize on surrounding farms. Application of manure produced noticeable differences in performance when compared to both maize grown without manure application and maize grown with chemical fertilizers.

The farm is also demonstrating soil conservation techniques, agroforestry and integration of crops and livestock. Basically, these are traditional practices and are therefore easily adopted. Ntonda is one area in Malawi which still has some forest areas and the people still place value on many indigenous and traditional systems of cultivation. We still use indigenous vegetables, indigenous fruits, insects for relish, and some roots and tubers for food. Traditional knowledge for increasing soil fertility includes the use of trees like "Msangu", under which crops are still grown. Vegetables grow for longer periods and onions grow better under these trees. Growing crops in the dambo areas, which have rich soil deposits, is another practice in Ntonda. People also plant vegetables, tomatoes and pumpkins in areas like anthills. This valuable traditional knowledge has rarely been promoted by government.

Each member of LOMADEF has set aside a plot in his or her own garden for activities similar to those being carried out on the demonstration farm. This has hastened the replication of the activities and organic methods in other community areas. Over a period of three years, workshops on organic farming have been organized at village level and in several church centres at the initiative of LOMADEF. These workshops deal with organic farming in general, but particularly with sustainable agriculture principles. The plan is to expand the activities by holding district workshops and using the previous participants as resource persons.

The demonstration farm also organizes field days for farmers and agricultural officials from all over the country. Government and NGOs involved in agriculture have brought more than 1,200 farmers on five different occasions to observe the benefits of organic farming and learn some simple organic technologies. The farm has been visited by both national and foreign dignitaries. All who have visited the farm leave with the impression that organic farming practices are an answer to the problems of resource-poor farmers. Visitors see that LOMADEF is a unique farm and that its activities are making positive contributions to the area. Many farmers are now practising the technologies observed at Lipangwe.

Two years ago, the LOMADEF group expanded its outreach by cultivating organic maize in various villages. There are currently about 40 families under this contract farming programme. The success of this programme encouraged us to enter into organic legume production. Twenty people benefited from this small project organized last year. None of them used chemical fertilizer or needed loans. LOMADEF shows that it is possible to use alternatives to conventional inputs and produce reasonable yields. Importantly, it has also improved the livelihood of the farming families in our programme.

Problems

Work on the demonstration farm has not proceeded without some problems. Labour supply, especially during peak periods, presents some difficulties. Cultivation on the farm is done solely by project members. At peak periods, such as weeding time, members dedicate all their time to their personal fields, which results in operational delays on the demonstration farm. In the dry season, the farm grows a variety of vegetables on irrigation. However, the water source quickly dries out. We are therefore in search of funds to dig a borehole to ensure a reliable water supply for growing vegetables. In order to extend the operations of the demonstration farm, we are also looking for funds to construct buildings in which to house and train farmers from distant places,

Conclusion

LOMADEF is dedicated to increasing organic farming in Malawi. The concept is not new, but it is not widely practised and at present there is no movement to specifically encourage organic farming. A good number of people, especially small-scale farmers, are becoming interested in the practice of organic farming. However, there is a need for encouragement, information sharing and more training. The people trained in organic techniques are spread thinly in the country and there is a need to cover wider areas to spread impact. Networking among interested individuals and the organizations they represent is needed, but funds are lacking to support these activities. The Lipangwe Organic Manure Demonstration Farm is very important in this process because it exposes small-scale farmers to low-cost agricultural technologies which everyone can easily afford. The smallholder farmers around the area are greatly attracted by the methods demonstrated on the farm, especially because agricultural input prices are so high. Our experience suggests that demonstrations managed by smallholders are attractive to other smallholder farmers. The demonstrations help to encourage farmers to adopt organic farming and low-cost technologies. In view of the importance of sustaining natural resources, especially the soil, and in view of the escalating costs of other farming inputs like chemical fertilizers and pesticides, organic farming has a big role to play in the agriculture of Malawi and most African countries. It is our hope that the experience gained at LOMADEF shows how we can organize to foster the concept of organic farming.

References

Government of Malawi. 1987: *Statement of Development Policies 1987 - 1996*. Zomba: Government Printer.

Ministry of Agriculture and Livestock Development, Malawi. 1994: *Guide to agricultural production* 1994/95 - 1996/97.

Ministry of Economic Planning & Development, Malawi. 1996: *Food Security & Nutrition Bulletin*. Vol. 7 No 1.

Chapter 26

Fambidzanai Permaculture Centre: Training for Ecologically Sound Agriculture in Zimbabwe

by Livai C. Matarirano

Introduction

Fambidzanai Permaculture Centre (FPC) is a trading trust located on a 40-hectare plot in Mt. Hampden, 21.5 kilometres west of Harare, Zimbabwe. It was established in 1988, largely in response to the ecological deterioration in the communal areas. *Fambidzanai* is a Shona word for communion, companionship or collaboration. In our case, it means mankind living in harmony with nature. FPC's mission is to enrich the quality of rural community life by offering participatory training programmes and demonstrating working examples of land use designs and related farm technologies. At FPC we believe that, in the long term, the only sustainable systems are those that provide for their own needs and do not exploit or pollute the environment.

What is Permaculture?

Permaculture, which is derived from the words permanent and agriculture, is a permanent, self-sustaining system of agriculture. It is designed to produce an efficient, low-maintenance, optimally productive integration of trees, plants, animals, natural and man-made features and human activities within a specific environment. Permaculture emphasizes the use of locally available resources (low external inputs) and little or no chemicals. With permaculture as a base, FPC seeks to use traditional and modern agricultural and technical knowledge to create systems that are socially acceptable, environmentally friendly and economically viable.

Courses

Fambidzanai emphasizes training in practical skills for land use design and planning, particularly skills that are appropriate and relevant to communal farmers. The training - which is targeted at farmers, development workers, government extension workers, college and university students and teachers - is through residential courses at the centre or outreach in the communities. This training results in improved living standards for the farmers as they develop useful skills for sustainable land use.

Some of the courses offered at FPC are:

- Introduction to Permaculture
- Practical Introduction to Sustainable Agriculture (PISA)
- Participatory Rural Appraisal
- Introduction to Natural Pest Management
- Natural Pest Management
- Organic Gardening
- Introduction to Sustainable Land Management
- Permaculture Land Use Design
- Holistic Decision Making
- Grazing Area Management
- Beekeeping
- Nursery Techniques
- Sustainable Dryland Cropping
- Permaculture Refresher
- Textile Screen Printing
- Thatching
- Brick and Production of Soil Stabilized Blocks
- Production of Micro-Cement Roofing Tiles
- Rammed Earth Building

There are one- and two-week courses, two- to six-month internships and two-year apprenticeships. Four trainers provide the theory and course guidance. External resource people are sometimes hired. The lands department provides back-up and supervises the practicals. Course fees for residential courses cover tuition, accommodations, meals, handouts and planting materials. At the current rate, a minimum of 15 trainees attending each course is needed in order for FPC to break even. The Centre can accommodate 25 participants at any one time.

PARTICIPANTS

Over the years, FPC has trained 1,166 people including farmers, development workers, government extension workers, students at college or university and teachers.

Table 26.1 Number of people trained at FPC per year, 1989-1996

Year	No. of trainees
1989	32
1990	89
1991	160
1992	151
1993	165
1994	117
1995	244
1996	208
Total	1166

While Zimbabwe is our main focus, some participants have come from other countries in Africa, Europe and America.

Table 26.2 Number of interns passing through FPC, 1991-1996

Country	No. of Interns
Australia	7
Botswana	1
Britain	12
Canada	5
Denmark	4
Germany	9
Kenya	5
The Netherlands	4
Sweden	4
South Africa	5
Swaziland	2
United States of America	10
Uganda	2
Zaire	1
Total	73

OTHER ACTIVITIES AT FPC

In addition to training, other activities at the centre include the production and sale of organic fruit, vegetables, herbs, field crops, small livestock, poultry, milk, honey, jam, indigenous and exotic seedlings, grinding, textiles, bricks, soil stabilized blocks, floor and roof tiles. FPC has 16 acres of arable land and raises livestock, which includes chickens, rabbits, ducks, geese, pigs and dairy and beef cattle. After supplying the centre's restaurant, surplus products from these units are sold.

As the centre is endowed with good clay soil, we run a building materials production unit. This unit produces bricks and tiles using manually operated machinery; it operates mainly in the dry season. When operating at full capacity, the unit can produce 3000 common bricks per day. Surplus materials beyond the centre's needs are sold. FPC also operates a textiles unit to generate income. This unit produces hand-printed textiles and garments using tie-and-dye, batik and T-shirt printing techniques.

News from the Field: The "multiplier effect" operational

- **Gwembe South Development Programme:** Recent articles from the *Agroforestry versus Land Degradation Newsletter* (1996, 1997) describe how Gwembe South has achieved positive results from Fambidzanai courses. In 1995, five people from Gwembe South, Zambia, attended the 2-week courses on Permaculture Land Use Design held at Fambidzanai. That training was considered very worthwhile. Encouraged by what was learnt, the Gwembe South Development Programme: Sustainable Agriculture Project decided to send a further seven extension staff and farmers to Fambidzanai during 1996, this time on specialized one-week courses. In July and August 1996, the five village Community Agriculture Workers and two GSDP Sustainable Agriculture field staff attended training courses on nursery techniques for plants and trees, and sustainable dryland cropping. Berda Moonga, one of the five CAWs who attended the nursery techniques course, returned to Kaumba village to help create a community nursery where they are raising both annual and perennial *Sesbania sesban* leguminous trees to help improve soil fertility in the farmers' fields.

Looking Ahead

- **FINESA: Reaching out to the communities:** Beginning in July 1997, the Farm Information Network for East and Southern Africa (FINESA), formerly the Farm Radio Network (East and Southern Africa), was slated to become an outreach extension organ of FPC. FINESA collects and disseminates tips on everything affecting the small farmer – food production, nutrition and health. The information is distributed as information sheets or scripts. Some of the scripts in recent packages include:

- use of earthworms to farm your garden
- good soil management increases harvests
- make compost pits in the garden
- more with mulch
- build and maintain contour ridges
- powdered cattle dung makes excellent fertilizer
- use good plants to control bad weeds
- knowing insect life cycles helps you control pests
- a simple fly trap
- good cow feed at the end of the dry season.

These are the technologies that FPC is demonstrating. Our problem has been reaching out to farmers who have trained at FPC and those who have never been to the centre. FINESA will make it possible to reach these people and meet their information needs.

- **Pelum College of Zimbabwe:** Graduates from training institutions in Zimbabwe have had problems settling down to their jobs in communal areas. Some non-governmental organizations in Zimbabwe are putting up a programme to redress this. Together with the PELUM (Participatory Ecological Land Use Management) Association and a number of other organizations – the Association of Zimbabwean Traditional Environmental Conservationists (AZTREC), the University of Zimbabwe Faculty of Agriculture, the Africa University Faculty of Agriculture and Natural Resources, the Nyuchi Dzakasimba, Chikukwa and Nyahode Union League Centre – FPC is setting up the Pelum College of Zimbabwe (PCZ). This "moving" college will train development workers in participatory development techniques and ecological land use management. Trainees will spend time at each institution covering relevant sections of the curriculum. At FPC, trainees will learn about permaculture and natural pest control. Current discussions are looking into funding and accreditation options for the college. If all goes well, the college will begin operating in September 1997.

References

Agroforestry versus Land Degradation, 1996: Newsletter No. 13, May. London: Methodist Relief and Development Fund/Africa Link, The Methodist Church.

Agroforestry versus Land Degradation, 1997: Newsletter No. 15, April. London: Methodist Relief and Development Fund/Africa Link, The Methodist Church.

CHAPTER 27

THE ANIMAL TRACTION NETWORK IN EASTERN AND SOUTHERN AFRICA

by T.E. Simalenga

INTRODUCTION

The use of animals for agricultural purposes in most of Eastern and Southern Africa was introduced more than 70 years ago, but widespread adoption did not occur until the 1940s and 1950s (Kjaerby, 1983; Kwiligwa, Shetto, Mkoga & Haule, 1992). In most of these countries, European settlers and missionaries introduced draft animals; oxen and donkeys were mainly used. South Africa is an exception. Animal traction has a long history there, dating back to before the arrival of European settlers in 1654. Animal traction was a prime source of power for both transport and agriculture in South Africa until the early 1920s, when fossil-fueled power gradually began to supersede animal traction on commercial farms (Starkey, 1995).

The spread of draft animals among African farmers depended almost exclusively on the use of the single-furrow mouldboard plough. In Ethiopia, however, a locally made wooden scratch plough (the "ard", locally known as the maresha) with horizontal fitted iron blades has been used widely for more than one thousand years. In Tanzania, the introduction of ox-ploughing has been linked with the introduction and development of profitable cash crops such as coffee, tea and cotton in the western parts of the country. Evidence from other African countries shows that cotton, groundnuts, maize, wheat and rice were the most important cash crops associated with the development of ox-ploughing. Some of the important examples are: groundnuts in Senegal and Gambia; cotton in Northern Nigeria, Niger, Uganda and Tanzania; Maize in Zimbabwe, Malawi and Tanzania (Kjaerby, 1983). The introduction of animal traction is clearly linked to the development of a commercialized economy where high prices on export and food crops accelerated the adoption of the technology.

Trends in Animal Traction Development

Over the past decades, there have been considerable efforts to develop and promote animal traction, supported by government, international agencies, non-governmental organisations and interested farmers. In Zambia, Kenya and Uganda, animal power development projects have been supported by the Netherlands government and the World Bank. In Tanzania, animal traction development has been supported by the Canadian International Development Agency (Mbeya Oxenization Project), in the Western Zone by the Netherlands government, and in Tanga, the Draft Animal Power project was supported by GTZ.

Various avenues have been pursued. Field performance and testing trials for animal-drawn equipment have been conducted at research stations and in on-farm trials (Ngendello, Well & Roeleveld,1995; Kwiligwa *et al.* 1992). Ploughs, harrows, seeders, inter-row and over- row cultivators, ox-carts, donkey carts, harnessing equipment for both oxen and donkeys and multi-purpose tool carriers have been designed and developed (e.g. Bukura MK 11 in Kenya, Kabanyolo in Uganda, donkey harnesses and carts in Kenya and Tanzania) (Mbugua, 1992; Akou, Afrida, & Odogola, 1990; Mkomwa, 1992; Kwiligwa, 1980; Oudman, 1992). Management, strategic supplementation and feeding levels of draft animals have been studied (Ndlovu, 1990; Aboud, Mutayoba & Mollel, 1997). Research and extension linkages have been developed, e.g. research and development of drawn weeders in Tanzania (Mkomwa, 1992; Shetto & Kwiligwa, 1992). Ox-training and farmers' training centres have been established. Local animal-drawn implement manufacturing companies and artisan/blacksmith development at village level have been promoted.

Despite all these efforts to develop and promote animal traction in the region, the rate of adoption of the technology among smallholder farmers is very low (Starkey, 1986; Mrema & Hatibu, 1990; Simalenga, 1994; Simalenga & Shetto, 1995). Human power is still the biggest power source used in crop production (Table 27.1).

Table 27.1 Comparison of power utilization in field operations (%)

AREA	HUMAN	ANIMAL	TRACTOR
North Africa	69	17	14
Sub-Saharan Africa	89	10	1
Asia (excl. China)	68	28	4
America	59	19	22
Overall	71	19	10

Source: Panin and Ellis Jones (1994)

The low rate of adoption can be explained by a number of factors. There is a lack of awareness and information to demonstrate the advantages and profitability of using animal draft power on farms. The equipment is not readily available in rural areas and, when it is available, it is too expensive to buy. Ownership of cattle or donkeys and access to draft animals is limited. There are taboos and cultural resistance to changes. There is lack of rigorous research and development in the design of appropriate animal-drawn implements to suite local conditions. There is lack of a participatory approach to extension and dissemination. Table 27.2 shows the potential for using cattle and donkeys in selected countries in the region.

Table 27.2 Estimates of the total number of employed cattle and donkeys and available power in six selected African countries:

Country	Cattle Population	Employed Cattle	Donkey Population	Employed Donkeys
Zambia	2,700,000	240,000	2,000	1,000
Zimbabwe	5,950,000	500,000	104,000	70,000
Malawi	1,400,000	70,000	2,000	1,000
Tanzania	12,000,000	800,000	750,000	200,000
Uganda	3,900,000	600,000	170,000	13,000
Kenya	12,000,000	700,000	30,000	20,000

Source: Dibbits, 1992

Key Issues in Promoting Animal Draft Power

Much information is available on the technical and economic advantages of animal traction development (Munzinger, 1982; Starkey, 1986; Panin & Ellis Jones, 1992). In recent years, promoters have viewed animal draft power as appropriate and sustainable technology for intensifying agricultural production and providing transport for rural communities. However, while it may seem a simple technology, its promotion and use can have complex social, economic and environmental consequences that are often not appreciated by animal traction promoters. Figure 27.1 attempts to summarise the complex situation under which animal draft power operates, its interrelationship with various disciplines and linkages of constraints that farmers face. The promotion of animal draft power requires the concerted efforts of animal scientists, veterinarians, agricultural engineers, planners and development workers. Networking among these scientists is one step to ensure that the existing information and experiences are shared to avoid repetition of mistakes and the introduction of inappropriate methods. The Animal Traction Network for East and Southern Africa (ATNESA) was launched to address this need and to facilitate the promotion of animal draft power through improved information exchange and cooperation in the region.

Figure 27.1 Animal draft power and its linkages at farm level

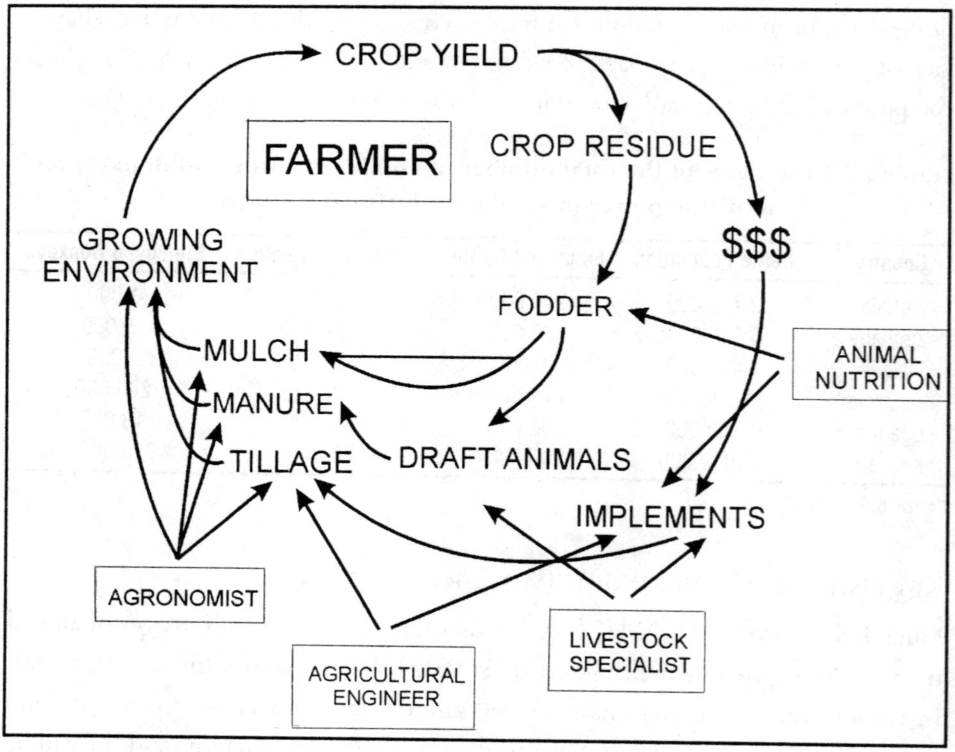

The Launch of ATNESA

In 1987, the Southern African Centre for co-operation in Agricultural Research (SACCAR) organised a regional animal traction workshop in Maputo, Mozambique. The workshop resolved that a regional information sharing network should be established under the auspices of SACCAR. However, there was no immediate follow-up to the idea. In November 1990, the AGROTEC (Agricultural Operations Technology) programme for smallholders in East and Southern Africa organised a course on planning integrated animal draft programmes in Harare, Zimbabwe. During the course, the decision was made to launch the regional network. A provisional steering committee of six people from different countries was selected to discuss organisation details and prepare the network's action plan. The first ATNESA major workshop was held in January, 1992 in Lusaka, Zambia with the theme "Improving Animal Traction Technology". A total of 107 people from 17 countries participated. During the workshop, two open general assembly meetings were convened. At the first meeting, organisational arrangements and draft statutes for the network were discussed and an ad hoc committee was nominated to finalize these. At the second meeting, the statutes were formally adopted and a new steering committee was elected. The steering committee was given the mandate to arrange a programme of activities based on the workshop recommendations.

Objectives and ATNESA Activities

ATNESA was launched with the main aim of improving information exchange and regional cooperation relating to animal draft power. The network aims to unite researchers, trainers, manufacturers, development workers, institutions and the users of animal traction in the region. Membership is open to all individuals and organisations interested in the objectives of ATNESA. These may include research institutions, development projects, NGOs, farmer associations and manufacturers. Organizations and individuals from outside the region are welcome as associate members. Currently 15 countries are participating members: Eritrea, Ethiopia, Kenya, Uganda, Sudan, Tanzania, Malawi, Zambia, Zimbabwe, Mozambique, Namibia, Botswana, Lesotho, Swaziland and South Africa.

ATNESA seeks to promote environmentally sound and sustainable systems using animal traction in the region; to promote a farming systems, farmerorientated approach to animal traction development and research; and to strengthen and promote links between individuals and organisations engaged in using draft animals and those conducting animal traction research, extension, training, manufacture and other related activities. It also seeks to stimulate and promote the dissemination and exchange of information, research findings and experiences in animal traction; to promote coordination and collaboration in development and research activities; and to encourage and stimulate the formation of national animal traction networks. In addition, it seeks to enhance the performance and professional development of researchers and technical personnel involved in animal traction activities through appropriate training initiatives; to facilitate assistance to network members on technical and organisational matters relating to the planning, preparation, funding, implementation and evaluation of their animal traction programmes; and to stimulate linkages between members and organisations capable of providing financial and/or technical support. Finally it seeks to organise regional activities, including biennial workshops, seminars, study tours and professional visits.

Based on these objectives, ATNESA has organised formal and informal national animal traction networks; facilitated thematic small workshops on specific themes, such as: gender issues in animal traction, animal-powered transport, weed control, donkey utilization, conservational tillage and environmental issues; and prepared an ATNESA directory of people and resource organisations involved in animal traction in the region. ATNESA has increased awareness of animal traction through formal publications, publicity materials and contributions to newsletters and journals; established a database with reliable statistical data on animal traction in the region; and promoted the exchange of information and networking through electronic e-mail systems.

The ATNESA Networking Model

Ideally, individuals or farmers and organisations involved in draft animal power or agricultural development would be well linked with colleagues involved in similar areas. However, in reality, this is not the case. Information flow and experience is restricted to single disciplines, limited geographical areas or research centres. Networks can help overcome such problems by providing a mechanism whereby people and organisations can exchange information and experiences or undertake joint activities. Networks can be classified by their membership, geographical scope, main activities and organisational structure. The steering committee of ATNESA decided to adopt an informal system of network organisation, based on national networks linked through a regional network steering committee. In this way, the secretariat responsibilities are delegated to members or national networks. Currently the ATNESA is coordinated by a steering committee of specialists from six African countries and representatives from two resource organisations. The steering committee is elected for three years. Elections take place during network general assembly meetings, which are held during major workshops.

PROGRESS AND ACHIEVEMENT

Workshops

ATNESA recognized that large, wide-ranging workshops were important for encouraging general information exchange and the cross-fertilization of ideas. However, smaller workshops on specific themes were also needed to allow specialists to concentrate on particular issues. Over the years, ATNESA has been able to arrange several important workshops: Improving animal traction (January, 1992); Gender issues in animal traction (May, 1992); Animal drawn carts (January, 1993); Weeding using animal power (November 1993); Meeting the challenges of animal traction (December 1995); and Donkey utilization and management (May, 1997). Through these workshops, more than 400 people from 35 different countries worldwide have been able to meet, exchange ideas and initiate collaborative research, training and extension programmes.

Formation of national networks:

ATNESA has endeavored to stimulate the organisation of informal or formal national animal traction networks in as many countries in the region as possible. The national networks have goals comparable to those of ATNESA, but they operate at national level. They generally aim to influence national policy and extension programmes in favour of animal traction. National networks have already been formalized in Kenya (KENDAT), Tanzania (TADAP), Ethiopia (ENAT), South Africa (SANAT) and Zimbabwe (APNEZ). In Zambia and Namibia, networking is assisted by national animal traction programmes. Discussions to start national networks are taking place in other countries.

Directories and databases

ATNESA members and national networks are compiling information on organisations and individuals involved in animal traction. In some countries, annotated directories of animal traction organisations and specialists have already been circulated (Simalenga & Hatibu 1993). Such information is used in regional mailing lists and will form the basis of the proposed ATNESA directory. Discussions are underway to establish a regional database on animal draft power and share information through the internet. A working group has been formed to look into the matter and suggest a modus operandi.

Publications

ATNESA, in collaboration with other organisations, is publishing a series of workshop proceedings and resource discussion papers, generally available to network members at no charge. The following publications are now available:

- Improving animal traction technology (Proceedings of 1992 ATNESA workshop)
- Gender issues in animal traction: a handbook (Proceedings of Gender workshop)
- Animal Drawn Carts: Guidelines for design, manufacture, testing and promotion (Published by ITDG based on animal drawn carts workshop)
- Meeting the challenges of animal traction (1995 workshop report)
- Bibliography on donkey utilization and management
- Animal power for weed control (in print)
- Donkeys for rural development (in print)

Some Lessons Learnt and Future Approach

It has been observed that networks can effectively link people who would not otherwise interact. The main benefits of networks are the exchange of experiences, information, materials/publications and cooperation in development and extension programmes. Networks can also stimulate and bring together funding and technical cooperation agencies and those in need of resources and support. ATNESA's experience suggests some lessons and constraints that should be considered by groups seeking to establish formal networks.

The need for a Secretariat

When ATNESA started, it had adopted an informal networking methodology which ensured that most of the work was delegated to the members and the national networks. Experience has shown that, to operate effectively, a network needs a small but effective secretariat or a coordinating unit, possibly with a coordinator and secretary. The major role of the secretariat is not to centralize the activities of the network, but rather to facilitate its execution. Currently, most of the coordination is done on a voluntary basis by the chairperson or the technical adviser; sometimes this has led to the delays of important activities and implementation of decisions taken by the steering committee. Efforts are now underway to solicit funds and establish an ATNESA office with a coordinator.

Funding

Networks need both user support and adequate resources for their activities. International networking is not possible without a constant flow of funds. ATNESA has survived on funds requested to carry out specific activities. Several institutions, donor agencies and regional projects, such as AGROTEC and now FARMESA (Farm-level Applied Research Methods for East and Southern Africa), have provided support from time to time. The future approach will be to establish a revolving fund through the sale of publications and possibly charge members a small fee for services rendered.

Communication

Communication is vital for successful networking. However, most of the countries where ATNESA operates have poor communication systems. This has sometimes frustrated networking efforts. Telephones, where available, are very expensive and letters take up to two weeks to be delivered.

Farmer involvement

While the objective is to be farmer-oriented, members are mainly salaried professionals. Farmers in the region have been involved in field visits during the workshops. We need to involve farmers in our networking activities right from the planning stage. Greater farmer involvement could come from farmer-based workshops and technology-transfer exchanges. The future challenge of ATNESA is to stimulate the developments of animal traction through participatory, farmer oriented programmes.

National networks constraints

Ideally, the management of national networks would be easily compared to a multi-national network. However, experience has shown that this is not always true. Lack of initiatives and network volunteers can cause the networks to perform below expectation. The main objectives of a national network, i.e. acting as a lobbying body to policy makers and providing networking information to end users, has not been fully realized to date. Lack of operational funds has resulted in a situation where most of the networks rely on the institutions where the office bearers are based, using the workplace resources for network operations.

Volunteer work

Most of the networking coordination is done voluntarily and this requires dedicated people who are motivated and rewarded through academic achievement. This is very difficult, especially in third world countries where the majority of people are struggling to make ends meet.

Collaborative initiatives

ATNESA has provided a platform for initiating joint research and development activities. Regional scientists have been able to conduct study tours, farmer exchange visits and plan for joint training programmes. Contacts for joint equipment manufacturing ventures have also been established through networking.

Conclusions

This paper has attempted give a brief overview on the trend of animal draft power development in the region. Despite the low performance and adoption rate of animal draft power over the years, it is quite clearly the most profitable farm mechanization option under smallholder farming systems and is increasingly supported by donors and governments as a means to meet the power demand of smallholder farmers To increase the awareness of the potential for animal traction in the region, ATNESA has evolved as an informational regional network focused on animal traction issues. The network has initially concentrated on providing a mechanism for information transfer and the exchange of experiences between members. This has been achieved through various publications and small regional workshops. Future activities will concentrate on undertaking joint activities in collaborative farmer training, research and development.

While the network is meant to ultimately benefit smallholder farmers, they have only been actively involved in the network through farm-based discussion during workshop planning and meetings. Future efforts will concentrate on ensuring that ATNESA members work closely with farmers to discuss draft animal power technology. It is hoped that animal traction workshops will be initiated specifically for farmers. It is concluded that the combination of necessity, enthusiasm, judicious planning and good fortune have allowed ATNESA to develop as an effective network that has made rapid progress. Whatever formal network organisational systems evolve, it is most important that active networking interactions be maintained. Networks can survive on little money, but do need adequate resources to sponsor regular sharing opportunities.

References

Aboud, A. A. O, Mutayoba, B.M. and Mollel, E. L. 1997: *Effect of plane nutrition and draft stress on body condition and levels of blood metabolites in donkeys in Tanzania.* Paper presented at the ATNESA workshop on donkey utilization and management, held in Ethiopia. May 1997.

Akou, A., Afrida., G. O. and Odogola, R. W. 1990: *The status of animal draft power in Uganda: Future trends and prospects.* In proceedings of a regional course on planning an integrated animal draft programme, held in Harare, 5-13 Nov. Bulletin No. 2, Harare: AGROTEC.

Dibbits H. 1992: *The role of animal power technology in Eastern and Southern Africa.* Paper presented at the 1st KENDAT conference held in Nairobi, 3-6 Nov. 11p.

Kjaerby, F. 1983: *Problems and contradictions in the development of ox-cultivation in Tanzania.* Research report No 66. Uppsala, Sweden: Scandinavian Institute of African Studies, 163p.

Kwiligwa, E. M. B, Shetto, R. M., Mkoga, Z. J. and Haule, S. 1992: Prospects of animal traction technology in Tanzania. In *Proc. of TSAE* 4: pp. 43-48.

Kwiligwa, E. M. B. 1980: *Evaluation of the U. A. C. tool carrier used in maize production system.* Unpubl. M. Sc. Thesis. University of Dar-es-Salaam.

Mbugua, J. K. 1992: *Design of draft animal technology, a bottom-up approach.* Paper presented at 1st KENDAT conference held in Nairobi. 3-6 Nov.

Munzinger, P. 1982: *Animal traction in Africa* Eschborn, Germany: GTZ.

Mkomwa. S. 1992: *Research and development experiences of MOP in animal draught technologies.* In Hatibu N. and Simalenga T. E. (eds.) Proceedings of National Animal Traction workshop held in Morogoro, 8-10 April. pp. 31-33.

Mrema, G, C. and Hatibu, N. 1990: Draft animal power and agricultural production: An engineering view point. In *Proc. of TVA,* 7: pp. 53-66.

Ndlovu, L. R. 1990: *Management of draught animals.* In proceedings of a regional course on planning an integrated animal draught programme, Harare, 5-13 Nov. Bulletin No 2 Harare: AGROTEC.

Ngendello, A., Well, E., and Roeleveld, L. 1995: *On-farm participatory research on the ox- drawn weeding technology in North-West Tanzania.* Paper presented at ATNESA workshop in Nairobi, Kenya, December.

Oudman, L. 1992: *The animal draught power development project in the department of Agricultural Engineering, University of Nairobi.* Paper presented at 1st KENDAT conference held in Nairobi, 3-6 November.

Panin, A. and Ellis-Jones, J. 1992: Increasing the profitability of draft animal power. In Starkey P., Mwenya, E. and Stares J. (eds.) *Improving animal traction technology.* Proceedings of 1st ATNESA workshop held 18-23 January, Lusaka, Zambia. pp 94-103.

Shetto, R. M, and Kwiligwa, E. M. 1992: *Animal traction research needs in Tanzania.* In proceedings of Animal Traction Workshop held in Morogoro, 8-10 April.

Simalenga, T. E. and Hatibu, N. 1993: *Animal Traction Directory for Tanzania.* ATNESA

Simalenga, T. E. (ed.) 1994: *Participatory research and development of agricultural engineering technologies.* Proc. of AGROTEC regional workshop on on-farm trials held in Embu, Kenya.

Simalenga, T. E. and Shetto, R. M. 1995: *Animal power for weed control: Experiences and challenges.* Paper presented at the 2nd ATNESA workshop held in Nairobi, Kenya.

Starkey, P. H. 1986: *Draught animal power in Africa: priorities for development, research and liaison*, Networking paper 14, Farming systems support project, Univ. of Florida, USA.

Starkey, P. H. 1992: *Networking for animal traction.* Network discussion paper 92/1. GTZ. Animal Traction Development, Reading, 44p.

Starkey. P. H. (ed.) 1995: *Animal traction in South Africa.* Empowering rural communities. Development Bank of Southern Africa. 160p.

INDEX